Hill Farms

A Volume in the Series
ENVIRONMENTAL HISTORY OF THE NORTHEAST

Edited by
Richard W. Judd and Ellen Stroud

Hill Farms

*Surviving Modern Times in
Early Twentieth-Century Vermont*

DONA BROWN

University of Massachusetts Press
AMHERST AND BOSTON

Copyright © 2025 by University of Massachusetts Press
All rights reserved

ISBN 978-1-62534-872-2 (paper); 873-9 (hardcover)

Designed by Deste Relyea
Set in Adobe Jenson Pro
Printed and bound by Books International, Inc.

Cover design by adam b. bohannon
Cover photo by Arthur Rothstein, *Farmers with scythes in a Windsor County field during the depression*, 1937. Courtesy Farm Security Administration.

Library of Congress Cataloging-in-Publication Data
A catalog record for this book is available from the Library of Congress.

British Library Cataloguing-in-Publication Data
A catalog record for this book is available from the British Library.

Contents

List of Illustrations vii
Preface ix
Acknowledgments xi
Guide to Pseudonyms xv

INTRODUCTION
1

CHAPTER 1
How Farming Came to Jamaica, and What it Became There
14

CHAPTER 2
How Jamaica Became a Problem
The Experts in the Hills
41

CHAPTER 3
The Eugenics Survey
A Diagnosis and a Proposed Solution
69

CHAPTER 4
What Farming Really Meant
99

CHAPTER 5
Standards of Living
127

CHAPTER 6
Things Left Unsaid
Family, Community, Work
152

CHAPTER 7
The Great Depression
The Problem Becomes a (Temporary) Solution
175

EPILOGUE
Returning to Jamaica
201

Notes 213
Index 255

List of Illustrations

FIGURES

FIGURE 1. "An Early Settler Clears a Homestead—1733" 18
FIGURE 2. "Height of Cultivation for Farm Crops—1830" 26
FIGURE 3. "Land Classes of Vermont" 55
FIGURE 4. "Farms Studied in Detail" 63
FIGURE 5. "Towns in Vermont Similar to the Towns Studied in Physiography, Topography, Percent of Land Area in Crop Land and in Size of Population" 67
FIGURE 6. "Pasture scenes showing an intrusion of weeds and trees on the cleared land" 101
FIGURE 7. "Farm Security Administration client with canned goods" 121
FIGURE 8. "Density of Relief" 191

TABLES

TABLE 1. Sheep in Jamaica, 1830–1880 23
TABLE 2. Rates of decline in sheep holdings, Jamaica and Vermont 24
TABLE 3. Distribution of sheep on Jamaica farms 25
TABLE 4. Cattle and sheep in Jamaica, compared with rates of deforestation and reforestation 27
TABLE 5. Reported reasons for staying in Jamaica, Waitsfield, and Cornwall 88
TABLE 6. Migration of native-born Jamaicans 91
TABLE 7. Origins of migrants to Jamaica 93
TABLE 8. Dairy farms, general farms, and self-sufficing farms in 1930 108
TABLE 9. Cattle holdings in three towns, 1930 109
TABLE 10. Farms using gross and net figures in three towns, 1930 113
TABLE 11. Farming for home use 120
TABLE 12. Comparative incomes, net and gross 128
TABLE 13. Family connections 140

Preface

As the story goes, rural New England has always been declining. The image of the abandoned farm has been stamped onto the history, culture, and mythology of the region since the Civil War and even before. Especially in the northern hills, the story of decline seems written into the landscape itself. In each generation, people left hill farms: first for the broad fields of the Midwest; later for jobs in the mills; later still, perhaps, just for the bright lights of the cities. Small towns dwindled; barns collapsed; pastures grew up to trees. And there is no doubt that much of that did actually happen. But from the very beginning, the story has taken on an aura of inevitability, as if predetermined by the irresistible natural forces of stony hills and cold winters—or by the almost equally irresistible impersonal forces of industrialization and modernization.

I have never really trusted that narrative. It seemed unlikely to me that the hill farmers in all those "declining" towns would have told their story in quite that way. But the voices of rural working people are rarely heard in historical sources. I could find no way to test my hunch until I came upon a trove of manuscripts from a highly improbable source: a collection of field notes compiled by a team of researchers working for Vermont's now infamous Eugenics Survey. Ironically, these researchers, too, were investigating rural decline. Over a three-month period in 1930, the team went from door to door in the small Vermont town of Jamaica, conducting extensive interviews with every one of the town's 158 households. In hand-written notes, they recorded the words of the people they interviewed, often verbatim. It is those voluminous notes that have enabled me to write this book.

By the time of that 1930 eugenics survey, most of Vermont's agricultural experts had long believed that successful modern farming demanded a total departure from older ways: greater specialization, bigger farms, more machinery. It seemed clear to those experts that the small, old-fashioned farms of the hills could never get big enough or rich enough to modernize. They were destined, instead, to fade away into the woods, to be replaced by state forests and second homes for vacationers. Yet the Jamaica manuscripts show that farms had been flourishing in the hills for a long time; they were not big, or rich, but they were successful in matters that counted to their owners. Some were still there in Jamaica in 1930, when the eugenicists came to town.

Life on those farms was no pastoral idyl, to be sure. Its cash rewards were small; it demanded a lot of work, and also a lot of what one sympathetic observer called "unenvious satisfaction with plain ways." But the Jamaica manuscripts cast the life of the hill farms in a different light. Above all, they reveal the remarkable resilience and proficiency of the people the eugenicists were studying: their capacity to make do with little; to work with the limits of their land; to adapt to a rapidly changing physical and social environment. The Jamaica records, read carefully, suggest that there was much that was lost when those hill farms disappeared.

These days, it has become increasingly clear that the path of progress laid out by early twentieth-century agricultural leaders has led northern farmers to a dead end. In 1930 there were some 35,000 farms in Vermont. In 2020 under 7000 remained. And even those farms that did get bigger, more industrialized and more mechanized, have found that it is not possible to get big enough or mechanized enough to satisfy global markets. Perhaps now, at last, the path of modernizing agriculture may be losing its aura of inevitability. For many years, the narrative of decline in the hills has fostered the belief in that inevitability. Minimizing or even erasing from the record the existence of thousands of small, old-fashioned, diversified farms, that old story has lent credibility to the idea that farms should exist only where efficiency can be maximized, and that successful farming depends on the relentless pursuit of growth at any cost. It is my hope that the real stories of Jamaica's farmers will lead readers of this book to question that narrative, and perhaps to rediscover some of the possibilities of a different path: limited, modest, and local.

Acknowledgments

To write this book, I have had to learn a great many new things. My training as a cultural historian, alas, taught me little about the qualities of Podunk Sandy Loam, the weight of sheep fleeces, or the number of potatoes in a bushel. Even more than usual, I have had to rely on the work of other historians and on communities of scholars.

The origins of this project date back to 2014, when I was invited to participate in a conference to celebrate the career of my former teacher Robert Gross, who was about to retire. The other speakers, a group of Bob's former students who had gone on to careers in all kinds of fields, made the perfect audience for that first exploration. (My talk that day, as I recall, began with an image of a farmer in my neighborhood rescuing her sheep from the floods of Hurricane Irene—in a kayak, one by one.) At that event, as always, I was inspired by Bob's generous and imaginative responses to my work.

The scholarly works of Christopher Clark, Richard Judd, and Brian Donahue have deeply influenced my understanding of both agricultural and environmental history, and how they enrich each other. I thank each of them for their support of my work. I am grateful to Katherine Jellison, Grey Osterud, and many others for welcoming me into the circle of scholars that make up the Agricultural History Society, and showing me, through their writing and conversation, what agricultural history can achieve. I am also fortunate to be part of a lively community of historians now working in Vermont. I have learned all kinds of important new things from the work of Deborah and Nicholas Clifford,

Mercedes DeGuardiola, Christopher Harris, Greg Joly, and Paul Searls. I have also benefited for many years from the existence of the Center for Research on Vermont, which has been an essential resource for students, faculty, and researchers across the state. I am grateful to its longtime director, Richard Watts, for keeping the Center afloat. And I thank Cheryl Morse, past director, for our thought-provoking conversations about landscape, farming, and love of the land.

I have also relied greatly on my students. Hannah Kirkpatrick was a wonderful research assistant, helping me slog through the process of digitizing copies of all those handwritten notes in the state archives. Several of my former graduate students have done pioneering work with the eugenics documents over the years. Nancy Gallagher introduced me to the material and then went on to write the first important book on the subject. In more recent years, Richard Witting and Tom Anderson-Monterosso both developed important projects of their own, and their work has played an important role in shaping my thinking about the role of eugenics in Vermont history. My colleagues in the History Department at the University of Vermont have been thanked many times over the years, but it bears repeating that together, they created and sustained something quite close to the perfect scholarly environment. I would like to thank Dave Massell in particular, this time, for his interest in the rural landscape of Vermont.

My debts to librarians and archivists have also been piling up for years. At the University of Vermont's Special Collections library, the entire staff is owed many thanks. Prudence Doherty worked especially closely with me, not only on my research, but also on a series of graduate seminars that taught me as much as they did the students. The staff of the Vermont State Archives is a remarkably knowledgeable, good-humored, and collegial group. I thank Mariessa Dobrick in particular for her patience and interest.

In Jamaica, I have also met with a warm welcome. The folks at the Jamaica Town Office took time from more pressing matters to assist me with boxes of rolled-up grand lists and big books of land records. Greg Joly offered me generous assistance in the early days of my work, introducing me to the town and letting me know that there was one person in particular whom I needed to get to know as soon as possible. It is not his fault that I was slow to follow up on his advice, but he was certainly correct to point me toward Karen Ameden, who possesses an extraordinarily deep knowledge of the town and its people. I am grateful to both Karen and Greg for their help, and especially for one afternoon's long and fascinating conversation about Jamaica's people and places.

The University of Massachusetts Press has been a great organization to work with, and I am delighted to have my work included in its distinguished environmental history series. Brian Halley has been a terrific editor, consistently supportive and providing just the right amount of nudging in what ended up being a very long writing process. I am also grateful for the incisive and thoughtful editorial comments of the two manuscript readers.

As always, I thank family and friends for their patience and support. My thanks go to Sarah first, because she made the biggest sacrifice, graciously giving up a favorite and very long-running joke by refraining from beginning every conversation with "Are you done with that book yet?" My dear companion Steve has, as always, been my one best advocate: a reliable and judicious critic, a fierce defender of my time, and a model as a scholar and a writer. The only thing I have to offer him in return for all that is a little more time to share now that the book is done.

Working on this book has reminded me often of the generations that came before me, especially my many relations who struggled with Texas fields as rocky as Green Mountain hill farms. Finally, though, it is the present day and its formidable challenges that have been most on my mind as I wrote. This book is dedicated to all the farmers who are seeking new agricultural pathways that are abundant, resilient, and just.

Guide to Pseudonyms

The records of the Eugenics Survey of Vermont are open to researchers, but the policy of the Vermont State Archives is that the names of the people who were studied by the eugenicists are not to be made public. In this book, I have assigned pseudonyms to individuals whose eugenics records are featured prominently. I have also used pseudonyms where the use of the real name would be directly traceable to another individual interviewed by the eugenicists. The interview number in each footnote indicates where the file can be found in the archives.

The following names are pseudonyms. If a name is not on this list, it is the real name of the person, and it indicates that the person was not interviewed by the eugenicists nor directly traceable to anyone in the eugenics records who appears in this book.

Ames	Graham	Morris
Andrews	Halvorson	Prince
Armstrong	Jensen	Randall
Baker	Kingston	Rider
Bauer	Leon	Rowland
Chipman	Livingston	Ryan
Clay	Mason	Smith
Dewey	Merrell	Tower

Hill Farms

Introduction

One afternoon in late fall, a woman visited a farmhouse in a small town in Vermont. The lady of the house welcomed her in and offered her a cup of tea. She had been expecting this visit. The two women talked for hours: about the family, the neighbors, and the household chores. The husband came in to meet the guest, and she inquired about his crops and land; the children made an appearance, and the guest asked about their work at school and their plans for the future.

But this was not just a social call: the visitor had a job to do. She went home with copious notes, tracking where the household members were born, where they had lived over the course of their lives, and their religion, ancestry, ethnicity, and educational attainments. She recorded the farm's acreage and how much it was worth; how much it produced and of what things; the income gained from the work of each member of the household. Making use of her own observational skills, she assessed the cleanliness of the house and dooryard, the home's modern conveniences (or lack of them) and its room decor. She even commented (one assumes in writing only) on the appearance and personalities of the family members and rated the family's social status.

Nor was the visitor operating alone: she was part of a team of three women who had been assigned to investigate the people of this town. The *Brattleboro Daily Reformer* reported on their presence in the polite language of the society column: "The Misses Anderson, Choate, and Rome . . . have been very much pleased with their reception . . . and deeply appreciate the courtesy which has been shown to them." As the correspondent explained it, the team was there to investigate "the effect of the moving of people from towns to cities." The results of the study would "point the way" toward an unspecified "constructive program for the state."[1] And indeed, after the team completed its work and returned to their office at the University of Vermont, they would calculate the data they had gathered and use it to produce a wide-ranging report about the deep-seated regional problem of "rural decline."

It was the little mountain town of Jamaica, Vermont, that welcomed "The Misses Anderson, Choate, and Rome" in the autumn of 1930. (There is no plausible explanation of how the town got that odd name.) But it might have been any one of dozens of other towns. Rural social surveys had become an increasingly popular tool of analysis for social scientists in the early part of the century. Their first users were church leaders concerned with rural depopulation and with the decline of rural institutions.[2] By the 1920s, the long-settled rural uplands of New York and New England were experiencing a flood of scientific inquiries into the causes of population loss, low agricultural productivity, and social decline. In 1928, the state of Vermont formed a statewide umbrella organization to coordinate and multiply the efforts of all those experts who were already in the field. The new state Commission on Country Life brought together a broad coalition of scientific, political, and social leaders to launch an unprecedentedly comprehensive investigation into rural population decline and farm loss. Over the next three years, the Commission's sixteen committees hired a bevy of researchers to survey every aspect of rural life in the state, from soil quality to tourist boardinghouses.

Our trio of researchers were thus part of that broad initiative; they worked for one of the committees, and they had already visited two other towns before going to Jamaica. But they were not trained as soil scientists, geographers, home economists, teachers, or clergy: they were eugenicists. Though it was technically an independent entity, the Vermont Commission on Country Life was in fact largely the creation of the University of Vermont's now infamous Eugenics Survey. Henry Perkins, University of Vermont zoology professor and the founder and director of the Eugenics Survey of Vermont, played a formative role in the creation of the Commission. Eugenical thinking influenced most of its committees, but the Eugenics Survey was officially represented on one of them, the Committee on the Human Factor. The field workers in Jamaica reported both to the Eugenics Survey itself and to the Committee on the Human Factor.

The story of Vermont's eugenics movement has been ably chronicled by other historians.[3] It goes almost without saying today that the research sponsored by the Eugenics Survey, like the work of eugenicists everywhere else, was in nearly every sense flawed, its impact profoundly destructive. But eugenical thinking had deep roots in Vermont's culture, and the Eugenics Survey had many supporters. Among the Eugenic Survey's best-known undertakings was its genealogical research into targeted families—research that was facilitated by allies in the state charities and aid organizations who forwarded the names of their clients to the eugenics investigators. The organization also worked with other supporters to create a broad network of new institutions to house children and adults with identified "deficiencies" ranging from supposedly inherited "pauperism" to the catch-all diagnosis of "feeblemindedness." And for Professor Perkins, the crowning achievement of the Eugenics Survey would be the passage in 1931 of legislation that authorized "voluntary" sterilization for the vulnerable institutionalized populations his work had helped to create.

But none of that explains why this team of eugenicists had been sent specifically to Jamaica. On the face of it, Jamaica was an odd choice. In Vermont, as elsewhere, the eugenics movement is best known for targeting the supposed genetic deficiencies of unpopular racial or ethnic minorities. Perkins himself spoke openly about his belief that people of French-Canadian descent, for example—the largest ethnic minority in Vermont at that time—were of inferior "stock" and a major cause of the problems he believed were plaguing Vermont.[4] But Jamaica was a quintessentially "Yankee" town, containing a relatively small number of French-Canadian and European immigrants. (Even the team's official report would acknowledge that fact: of the three towns they studied, Jamaica had experienced the "least infiltration of stock."[5]) Most residents carried exactly the kinds of "good" genes ordinarily praised by eugenicists; indeed, many were descended directly from the venerated "pioneer" settlers of the region.

In fact, however, it was just these kinds of "Yankees" who emerged as the real problem race in most of the Vermont eugenics studies. Almost all the subjects of the Vermont eugenics studies were "white" and even "Anglo-Saxon." And the fear of a degenerating white race was actually a key feature of eugenicist thought everywhere. (The infamous Supreme Court case *Buck v. Bell* [1927], for example, concerned the involuntary sterilization of a young white woman in Virginia.) For Vermont's eugenicists, it was the deterioration of the state's rural Yankee "stock" that posed the most profound threat to the future of the region.[6] Still, the eugenicists did not label this study "Degenerate Yankee Families in Rural Vermont," or recommend sterilization for the inhabitants of the town. Nor were they attempting (this time) to root out "dependent, defective, and delinquent" families from otherwise thriving communities. The Jamaica study was a different kind of project. It was known simply as the "Migration Study."

Eugenicists believed (in common with many others in the 1920s) that rural decline in Vermont was caused chiefly by the migration of the so-called "best and the brightest" away from the farms and villages, a migration that left communities socially and genetically depleted. That kind of rural decline was a problem closely associated with Vermont's so-called "hill farms"—a term that denoted much more than geography. Most of Vermont's land is hilly, but not all farms were "hill farms." In an earlier time, the term had been used simply to refer to farms that were small, impoverished, and remote—often, but not always, on higher ground. Over time, it had come to acquire more negative associations with isolation or backwardness—a little like the word "hillbilly," that derisive term for the white but poor highland farmers of the southern Appalachians. A "hill farmer," in that newer sense, was someone who was not changing with the times.

By the early twentieth century, a label like that was needed, because so many other farms in Vermont *were* changing, turning into efficient modern businesses that produced dairy products, eggs, apples, and maple syrup for distant urban markets. As railroads and then auto roads transformed the relationships of farms to markets in the northeast, Vermont farmers in favorable locations were shifting away from the old mixed farming system they had grown up with. Producing for distant urban markets drove changes to their farming practices, in everything from accounting to the use of artificial fertilizer. Encouraged and instructed by a new generation of agricultural specialists, those farmers learned to take risks and borrow money. "Hill farms," in contrast, were by definition far from markets. Their operators, unable or perhaps unwilling to modernize, persisted in their small-scale, diversified, and (so it was thought) only modestly productive farming practices.

A generation before the eugenics study came to Jamaica, the hill farms were already losing population, but they had still possessed tremendous cultural resonance as living reminders of the state's rugged rural heritage and hardy, plain-living ways. That emotional weight had practical implications. The leading agricultural authorities in Vermont had once believed that it was their duty to support hill farms and their inhabitants, and to defend the hill towns from the demographic forces that threatened to destroy them. As one member of the Vermont Board of Agriculture put the case at an 1883 meeting, "Vermont is preeminently a 'hill-country' and all questions involving the consideration of agriculture in any form must recognize this peculiarity," adding that "If . . . our hill farms can be made to pay, . . . the future of successful agriculture in Vermont is assured."[7] It was only gradually that this conviction was undermined by new ways of understanding the nature of soil, of farming, even of human beings.

To be sure, even in the 1920s—in the "high unpleasant noon of Coolidge prosperity," as one critic later described the decade—many Vermonters still had fond associations with those old-fashioned farms.[8] But by then, increasing numbers of Vermont's leaders were also associating hill farms with the intractable social problems that had confronted the state for decades: the continuing population loss, farm "abandonment," and statewide economic and demographic stagnation. These were problems shared with many other rural places, but in Vermont—of all the New England states the most committed to agriculture—the future well-being of the state seemed uniquely tied to its agricultural prosperity, and the hill farms were increasingly seen as a barrier to that prosperity. By the late 1920s, things had reached a point where policymakers were beginning to consider the idea—self-defeating as it may now seem—of "solving" the problem of rural decline by removing the farmers from the hill farms.[9]

Such was the logic that shaped the eugenics survey of Jamaica. Three towns with rapid population loss were chosen for study. After spending the spring working in the town of Cornwall (which they expected to be in "fair" condition) and the summer in Waitsfield (in "good" condition), the eugenics team spent the autumn in Jamaica, which they anticipated would be in "poor" condition. They visited every one of the 158 households in town, carrying with them a set of pre-printed blank forms that required them to enter a three-generational chart of the family; the size of the household; and the occupations, birthplaces, religion, and yearly income of its members. There were places to note the acreage, the crops, and the number of sheep and cows the family owned. Larger blank areas required the interviewers to record the interviewees' "reasons for coming or staying" in town; the "condition of the farm," the "condition of the home," and the family's "social status." A half-page of empty space labeled "Notes" would be filled with long and sometimes seemingly verbatim reports of the interviewers' conversations with the residents. As a resource, these field notes are a treasure trove of information both qualitative and quantitative.

Of course, there are pitfalls in using materials like these. The eugenics records present the usual difficulties encountered by historians who make use of records generated by powerful and hostile institutions.[10] The sheer fact that the investigators were searching for signs of genetic "decline" in the population complicates any attempt to draw useful evidence about their subjects from their work. Like other materials created by such unsympathetic observers, these records must be read "against the grain." The three members of the eugenics team shared a worldview not only with other eugenicists, but with a broad network of scientists and social scientists that emerged in the early twentieth century—the very experts who now viewed places like Jamaica as the cause of the state's problems. And beyond that were still more unavoidable difficulties. The team's interpretation of what they saw in Jamaica was shaped by deep, scarcely conscious assumptions about social class, gender, and race, as well as by their own personal areas of ignorance and expertise. They seemed to know little about farming and less about working people. Their judgments were often heavy-handed, and their interpretations sometimes patently incorrect.

Even so, the documents are of great value—and not simply as a lesson in how not to conduct scientific research (although they are certainly instructive that way, too). The three women were diligent recorders, and they carefully documented many things they did not fully understand. Not infrequently, their own field notes belie the judgments they subsequently recorded, inadvertently revealing a community and its people far different from what they expected or even recognized. The very fact that the eugenicists felt so free to be candid in their notes—which they clearly considered private—offers a critical advantage to the researcher. And at key moments, the eugenicists' records afford almost unfiltered access to the residents' words. They did attempt to guard the anonymity of their informants: they assigned pseudonyms to the town and its villages, and they did not use individual names in their final report (although at the time, it would have been easy for Jamaica residents to recognize themselves and each other). Still, from the standpoint of historians (though surely not for the families who were investigated), it is a great stroke of luck that all these handwritten field notes managed to survive, and to land in a library where they would be open to public scrutiny.[11]

The members of the Vermont Commission on Country Life and the experts they hired left copious paper trails detailing their research processes and their judgments about rural life. But even in that age of radio and mass print, the residents of Jamaica and other hill towns were what social historians used to call the "inarticulate"—a silly, inaccurate word, but one that does make plain the glaring absence of their words and ideas in the historical record. The field notes thus offer a very rare opportunity to hear from rural people themselves about their land and their lives, an opportunity that is particularly valuable in the context of a time when so many other people were airing their ideas about farming.

And Jamaicans had a lot to say to the eugenicists. (Those of them who were interviewed were of course the people who had stayed in town or returned there.

It seems likely that many who left town and never returned would have told very different stories, but they were mostly unavailable to the investigators.) The fall of 1930 was a difficult time in Jamaica: the population of the town had reached its lowest point, little more than a third the size it had been in 1850. Most of its factories were gone, many of its fields had already grown back to woods. One entire hamlet had been emptied out after timber investors bought the land it stood on. All this at a point when the slow-moving catastrophe of the Great Depression was yet to make its full impact on the region. But most of the Jamaica residents encountered in the field notes do not sound defeated. (Not that they had no complaints at all: they frequently brought up the worrisome shortage of jobs in the area.) By turns they could be outspoken or laconic, worried or confident, judgmental about their neighbors or sympathetic to their problems. One purpose I have in writing this book is to allow those voices to be heard.

I hope that my book will thus achieve two different goals. The first is to tell the story of a long conflict over ideas about farming, and in a larger sense over the proper relationship of humans to the land. That story is in many ways not a new one. It fits within the established historical narrative of the great transformative processes that overtook northern American agriculture in the nineteenth and twentieth centuries. And it deals with a critical moment in that process, one that has been explored in a number of historical accounts of the 1920s and 1930s.[12] But my story also complicates the broad account (by turns triumphalist or tragic) of commercialization, modernization, and industrialization. It does so by focusing on the local and the particular.[13] It is certainly correct to say that farming in Vermont (and everywhere else) was transformed—commercialized, modernized, centralized, and industrialized—and that most of these transformations took place in the years discussed in this book. But Jamaica's story also highlights how slow, uneven, painful, contradictory, and inconclusive those processes could be.

The story is worth telling on its own, but it also serves a second purpose: to provide the framework that allows for a deeper and more challenging exploration of how the people of the hill farms themselves viewed their predicament—in particular, what they thought about the problems the experts diagnosed in their soil, their agricultural practices, and their community. By placing the words and ideas of Jamaicans side by side with those of the advocates of modernization, I believe I have uncovered in the field notes an insider's view of these transformations.

The first three chapters of the book offer three different accounts of how the characters arrived at the point of their encounter in the fall of 1930. Chapter 1 tells the story of how New England farmers came in the first place to settle towns so high up in the Green Mountains and how they managed there. Against older historiographical accounts that portray the move to the hills as some kind of mistake, I have followed the trail laid out by historian Brian Donahue in his investigation of agriculture in eighteenth-century Concord, Massachusetts. His investigation begins with the simple premise that Concord's farmers (and here, Jamaica's) knew what they were doing.[14]

At first, farming in the mountains was not unlike farming downhill. Like other New Englanders, Jamaica farmers carried on with the system they had inherited: diversifying their production, raising crops for market while also supplying their own household needs. But by the time the eugenicists came to Jamaica in 1930, five generations of farmers had been living there—long enough to learn to adapt to the opportunities and limitations of their corner of the world. Hill farmers like those in Jamaica assessed market risks and opportunities by their own careful calculus. No market opportunities escaped their notice, but they most often took a cautious and self-limiting approach. By the end of the nineteenth century, their continued adherence to older safety-first practices was beginning to seem increasingly out of step with their peers in the valleys. (All of that was implied in the term "hill farm.") Yet that would remain the farming system passed down all the way to the Jamaican residents who encountered the eugenicists in 1930—a little the worse for wear, but still recognizable.

Chapter 2 recounts how Vermont's politicians and agricultural experts came to view hill farms as a problem, and how they responded to that problem. The influence of new sciences and social theories reshaped popular thinking about farming in general, and about hill farms in particular. By the 1920s, those modernizing impulses were having an impact even on the typically cautious leaders of Vermont. Indeed, that is what led to the creation of the Vermont Commission on Country Life and its many surveys and inquiries—an unusual expenditure of money and time for the pay-as-you-go managers of the state. In her book *Every Farm a Factory: The Industrial Ideal in American Farming* (2003), Deborah Fitzgerald argues convincingly that American farming was systematically industrialized in the early twentieth century, not only by new machines and techniques, but by a comprehensive framework of "industrial logic"—ideas developed and promoted by new policy experts, agricultural economists, bankers, and engineers. Vermont was not in the vanguard of industrialization of farming, to be sure; too many structural and environmental factors still lay in the way. But the logic Fitzgerald described worked its way through Vermont, as it did through almost every corner of the nation.

Chapter 3 turns to the eugenics side of the story, and particularly how it was shaped by its lead investigator, Elin Anderson, who was tasked with diagnosing Jamaica's problems and proposing how to solve them. Anderson interpreted the data her team collected from Jamaica in a rather surprising way—surprising at least for a eugenicist. In the last analysis, she concluded, it was not the residents' genetic inferiority that was causing the town's decline; nor was it because "the best and the brightest" had fled. Instead, Anderson argued, it was the social environment of the hill towns that was weakening the townspeople's native strength of character. Perhaps that interpretation may seem a little more compassionate than the often-brutal judgments of other eugenicist investigators; at least the people of Jamaica were cleared of the charge of racial deterioration. (That was characteristic of Anderson's work as a eugenicist: her crowning achievement would be the influential study *We Americans*, a positive portrayal of the ethnic diversity of Burlington, Vermont.) But as it turned out, Anderson's report to the Commission on Country Life played a critical role in supporting a radical proposal for the comprehensive removal of people from the hills.

These, then, were the paths that brought the people in this story into contact with one another in the fall of 1930. The second half of the book explores what Jamaicans said to the interviewers. The next three chapters can be understood as a kind of rejoinder to Elin Anderson's analysis, a rejoinder drawn mostly from the eugenics team's own documents. Here I have marshaled other kinds of evidence—material mostly unavailable to (or simply not used by) the eugenicists—to interpret the interviews from several different angles. Chapter 4 examines the central question of farming itself. Exactly how were Jamaicans using their land in 1930, and why?

On this subject, perhaps more than any other, Jamaica's farmers articulated perspectives that contrasted strongly with those of the experts. Under increasing pressure from several directions, they held on to what they could of the agricultural practices they had inherited. Keeping old habits no longer approved by agricultural authorities, they grazed their animals in the woods, paid little attention to the upkeep of their pastures, let parts of their fields go back to brush, and maintained an emphasis on farming for household subsistence. Against expert advice, too, they worked longer hours on jobs away from the farm. Farming fewer and fewer acres of their land, they nevertheless held on to the security of subsistence production—and to the name and identity of "farmer."

In *This Land, This Nation*, historian Sarah T. Phillips observes that twentieth-century agricultural experts' devotion to measuring farm efficiency "left little room in the moral imagination for farmers who might have willingly chosen a lifestyle apart from profit maximization."[15] Chapters 5 and 6 examine whether some Jamaica farmers might have done just that: chosen a way of life "apart from profit maximization." Chapter 5 addresses the question of the material poverty of the hill towns. Most outside observers believed that those towns suffered from an exceptionally low "standard of living," a term newly in vogue among social scientists. If low incomes and a lack of modern conveniences are any indication, there was certainly poverty in Jamaica. But the findings of other contemporaneous studies—especially those of a group of home economists at the University of Vermont's own Agricultural Experiment Station—paint a more complicated picture. The eugenicists believed that the best strategy for hill town residents was to move somewhere else, to a place with better opportunities to get ahead. In reality, that path was seldom clear, and in any case, it did not often lead where the eugenicists thought it would. Chapter 5 thus explores the real material advantages and disadvantages of staying in town, based on the information Jamaicans gave to the investigators.

Chapter 6 raises a related but even more problematic question: What did Jamaicans say about the *non*-economic reasons they stayed in town? What was the root of their attachment? At a time when so many had already gone, what prompted some still to remain, in Hal Barron's memorable phrase, among "those who stayed behind"? This type of question, about the inner lives of rural people—or about anybody who has left few written records—is tricky.[16] It would be easy to fall into the same trap the eugenicists did (it seems so obvious now) and read one's own feelings and beliefs about rural life into the record. Most Jamaicans, generous as they were with their responses to questions about income and farming practices, appear to have been less inclined to reveal to the eugenicists their feelings about such intangibles as family ties, community cohesion, meaningful work, and personal autonomy. Their answers to direct questions were usually clear, but they were also very brief: *It's home*; *My people are here*; *More freedom here*. And often, the important questions were neither asked nor answered. This chapter necessarily includes a few more "perhapses" and "maybes" than does the rest of the book, but I have attempted to illuminate those brief phrases with the help of other local sources.

One thing is plain enough: the reasons the residents of Jamaica gave for wanting to stay were sound ones, ranging from practical considerations—an inexpensive farm, a relative's assistance—to the most personal and non-material of satisfactions. Whatever they "ought" to have done in response to the declining opportunities in the hills, a significant number of residents stayed home because of particular features they valued about life in Jamaica: a dense network of affiliations; the respect accruing to long family residence in a small town; and, for the men at least, a more varied work life that included some measure of autonomy. With such matters at stake, it is not so surprising that some residents chose to "stay behind" as long as they could.

As it happened, the Great Depression of the 1930s unsettled for a time the certainties of the experts. Chapter 7 explores how that happened. Even today, northern New Englanders sometimes repeat a well-worn joke: their farming grandparents and great-grandparents had not suffered during the Depression—because they had been "too poor to notice."[17] The joke ostensibly makes one kind of point: that the people on those farms were so poor they would not miss the radios and electric lights they had never had anyway. But the underlying premise of the joke slyly suggests a different point: anyone who could get by without *noticing* the Depression was clearly in possession of something worth more than cars or radios. Even in bad times—perhaps especially in bad times—the joke implies that hill farms were uniquely able to provide basic security, decent living, and even self-respect when those things were in short supply elsewhere.

The Great Depression was a profoundly destabilizing experience for rural and urban people alike, and for social scientists as much as for the farmers they studied. It undermined many people's belief in progress, even in modernity itself. As a result, for a time the Depression overturned many certainties about rural depopulation and its solutions. Experts, many of them now associated with the New Deal, continued to study the problems of hill farms during the Great Depression. Now, though, the whole question was tangled up with serious doubts about the survival of capitalism and the political stability of the nation itself.

Unsurprisingly, then, the Depression generated contradictory ideas about how to tackle rural problems. On the one hand, the New Deal swept into power the most forward-looking scientists and social scientists in the nation: dedicated to modernizing agriculture and weeding out farms that could not compete. When they looked at hill farms, whether in Vermont or in Appalachia or the Ozarks, they saw poor people, plain and simple, deprived of indoor toilets and electric milking machines. But at the same time, the Great Depression also raised up many defenders of what popular writers at the time called "farming as a way of life:" old-fashioned, self-sufficient, a bulwark against the ravages of economic catastrophe. Thus, the Depression set the stage for a many-sided debate about why, whether, and how to preserve places like the hill farms of Vermont. That debate unfolded in many parts of the country where federal aid for farmers was now on offer. But northern New England, and Vermont in particular, played a role far beyond its political (or agricultural) importance.

Within the state, a group of notable Vermonters now took up the cause of the hill farms, seeing in them a kind of self-sufficiency and stability that the rest of the country sorely needed. And a surprising number of those defenders of the hill farms lived near (or actually in) Jamaica. Some had been born in the area. George Aiken, perhaps the most admired of all Vermont's twentieth-century politicians, hailed from Putney, just two towns downhill from Jamaica. Aiken was elected governor of Vermont in 1936 and served as US senator from 1940 to 1975, but back when he was the state's lieutenant governor, it was chiefly his influence that would scuttle an ambitious New Deal plan to buy out and relocate the owners of the so-called "submarginal" hill farms.[18] Aiken's 1938 essay, "Not So Submarginal," was part of his book *Speaking from Vermont*, a kind of intellectual campaign biography written for his senate race. It was also a memorable defense of the hill farms and their importance to Vermont.

On the other side of the Green Mountains lived author Dorothy Canfield Fisher, who for decades had been celebrating the self-reliance and independence of her hill farm neighbors in every literary setting she could command. In her introduction to the 1937 Federal Writer's Project's guide to Vermont, she made the tongue-in-cheek proposal that Vermont be designated as a kind of national park for safeguarding old-fashioned farms and ways of living.[19] By then, she was no longer alone in her admiration for the hill farms and their way of life. Other writers—including many new migrants to the state—were coming to the rhetorical defense of the farms. Often they were moving to the hills and acquiring farms of their own.

Joining the ranks of back-to-the-landers escaping from the Depression was the radical economist Scott Nearing, who moved to Jamaica itself with Helen Knothe in 1932. From their new mountain stronghold, the couple wrote a series of back-to-the-land books advocating simplicity and subsistence farming as a solution to the problem of earning a living in a hostile world. The Nearings' writing did not appear in the mainstream press until the 1950s, and even then, did not find much of an audience until the republication of their books in the 1970s, when they became key texts of a revived back-to-the-land movement. In the meantime, however, the Nearings' underground reputation spread, and they played an important role in attracting new migrants to the sparsely populated hills on the west side of Jamaica.

Indeed, at the height of the Depression, a map of Vermont literary and artistic figures might have made it seem that Jamaica was the state's cultural epicenter. Dozens of writers and artists clustered around the town, each in their own way championing Vermont's traditional farms. They included poet Walter Hard in Manchester; in Dorset, novelist Zephine Humphrey and her husband, landscape painter Wallace Fahnestock; writer Frederick Van de Water in Dummerston; agricultural journalist Charles Morrow Wilson in Westminster; editor Vrest Orton in Weston.

But perhaps the best-known literary figure to be identified closely with the hill farms of northern New England was Robert Frost. Frost owned several farms over the course of his life, but his longest association was with the town of Ripton, Vermont, a mountain community that had lost even more of its population than Jamaica did. Frost's experience with farming in the northern hills informed the imagery of many of his poems, and several of them contained explicit defenses of the kind of small-scale, diversified, semi-subsistence farming commonly practiced in Jamaica, Ripton, and other hill towns. One major work was called "Build Soil." Written in 1932, it included some old-fashioned advice for the farmer on a "run-out mountain farm." Do not waste too much time on market crops: "[N]ot for a long, long time./But what you raise or grow/Why [,] feed it out/Eat it or plow it under where it stands/To build the soil." That is to say, provide first for your own needs: nourish the animals ("feed it out"), or the people ("eat it"), or the land itself ("to build the soil").

In 1932, Robert Frost would probably still have been considered a visitor by most Vermonters. (He owned a farm in Shaftsbury, to be sure—but he was there mostly in the summers.) Still, in this passage, he delivered a message any native-born Yankee hill farmer could have endorsed: "The moral is, make a late start to market."[20]

CHAPTER 1

How Farming Came to Jamaica, and What It Became There

The town of Jamaica is almost—but not quite—as far up in the mountains as a Vermont town can be. The main ridgeline of the Green Mountains runs ten or fifteen miles to its west through Stratton and Winhall, towns now largely subsumed by the Green Mountain National Forest and the Stratton Mountain ski resort complex. Jamaica's landscape is only a little less dramatic than that of its most elevated neighbors: a jumble of hills, peaks, ledges, and outcroppings separates its human settlements into a patchwork of isolated villages and hamlets. The roads follow the winding paths of mountain streams, mostly rocky and shallow in dry spells but prone to dramatic flooding from heavy rains or melting snow. Contemplating the heavily forested hillsides today, it is difficult to imagine how people ever thought they could farm here.

The first agricultural historians of northern New England, in fact, sometimes argued that it was all a mistake. As they told the story, migrants from southern New England, already wearing out the fertility of their land, were careless in their assessment of the northern hills. They neglected to consider proximity to markets, underestimated the severity of the climate, or were deceived by the short burst of fertility that followed the initial clearing of the land.[1] More recently, though, agricultural historians have been making the case that farmers like these might actually have known what they were doing.[2]

In many ways, Jamaica's story was typical of many other upland northern communities, settled nearly at the end of the "hiving" process that sent southern New Englanders west and north from their coastal beginnings. But that was not primarily because they feared the mountainous landscape. Massachusetts colonizers had other things to fear when they attempted to push the frontier boundary north. A long series of wars with the aboriginal Abenaki inhabitants of the region and their French and Mohawk allies prevented most English settlement in what is now Vermont until after 1760, when the surrender of the French imperial forces fatally weakened the Indigenous alliance. Even after the English victory opened the land to settlement, disputed land titles granted by competing interests in New Hampshire and New York sparked a conflict that slowed migration for years more. When Jamaica was finally organized in 1781, its charter was granted by a newly formed Republic of Vermont, technically still in rebellion against New York and unrecognized by the United States.

Clearly, moving north was a calculated risk. The land was inexpensive (in part because of those contested titles), but it would require hard traveling to get there and hard labor to transform it into farms. Still, there were good reasons for migrants to settle there. To begin with, they were propelled north by a strong "push" factor: population increase, combined with the constraints of southern New England land systems. Widespread land ownership was fundamental to the social structure of the New England colonies, underwriting as it did both political and family stability. No matter how difficult the terrain, the northern settlements offered the hope of retaining or reclaiming the security and social status that land ownership provided.[3] Many things would change in the years to come, but the descendants of those first Jamaicans would continue to value their farms for similar reasons.

At the same time, many aspects of the upland physical environment were actively appealing to migrants. There were practical reasons to prefer high ground. The lighter and thinner soils of the hills were easier to work than the frequently flooded clay soils of the valleys. Hillsides experienced fewer late frosts, and sloping fields drained more quickly in spring.[4] Carefully sited farms could take advantage of the increased sunlight at higher elevations: one archaeological study found, for example, that more than four-fifths of the earliest identifiable Green Mountain farmsteads were situated on slopes facing south and southeast.[5]

Evidence from the settlement period suggests, too, that early migrants understood that the first high yields after deforestation were a one-time-only bonus. One report from 1809 summarized the common understanding: for a few years after the forest was cleared, the land would bear "corn and other kinds of grain, in large quantities." But even though that first flush of fertility would soon subside, afterward the land would "naturally" become "rich pasture or mowing."[6] Agriculture promoters in Vermont a hundred years later had not yet forgotten what had drawn their ancestors to the hills. As one man explained at a Vermont Board of Agriculture meeting in 1884, "the first settlers looked for places with the quickest returns: This took them to the hills, on which a crop of grain might be grown the first year after the removal of the timber." And even though "the soil might prove to be rocky and shallow, still it served to produce good crops for many years."[7]

Leaving Petersham

One family's story illustrates how such calculations drew people north. The three Chase brothers—the youngest sons of a family of fourteen children—moved north to Jamaica from Petersham, Massachusetts, just a few years after the new town's proprietors had made their first distribution of lots. The town the Chase brothers left behind, like the one to which they were migrating, had been among the last in its own area to be settled by Europeans. Petersham was located high on a ridge in the hills of north central Massachusetts—the kind of place New Englanders had avoided for as long as there was coastal and river valley land to be had. That land had first fallen under English control back in the 1670s when its Nipmuc inhabitants were defeated in King Philip's War, but it was not settled by New Englanders until decades later in the mid-1700s, when the colony of Massachusetts granted the land to veterans of the long wars against the Abenakis and their allies.[8] The Chase family acquired land in Petersham in 1750, but it did not take long for the sons and daughters of the household to run up against the limits of the place. For their generation, the north country—however remote and mountainous—would be the land of opportunity. Of the thirteen Chase siblings who lived to grow up, five migrated to Vermont, one to New Hampshire, and three to upstate New York.

Elisha, Stephen, and Peter Chase bought land in the new town of Jamaica, but they did not actually buy farms. In the northern settlements, the New England system of town planting finally broke down completely. The carefully laid out allotments shaped by earlier southern New England proprietors—men who had at least seen the land they were dividing—were replaced by the speculator's grid, property lines cutting across hilltops, wetlands, and brooks without regard to the terrain or resources of the land.[9] Those who actually wanted working farms, rather than speculative investments, had to buy, sell, and trade their way to plots that corresponded more nearly with the lay of the land.

The Chase brothers first appeared in Jamaica land records in 1783, when Elisha Chase, the oldest of them, bought land from one of the original proprietors.[10] It was another ten years before middle brother Stephen made a separate purchase of land, and the youngest brother, Peter, recorded his first land purchase six years after that.[11] It took until the 1810s for the brothers to piece together three separate farms. The buying and selling continued even then, as they traded parcels with each other and with their neighbors.[12] Stephen died in 1824, and Elisha in 1828, but Peter lived until 1851. Over time he and his wife Polly accumulated a good property on the south side of South Hill, sloping down toward the brook called, at different times, Wardsboro, Whetstone, and Mill Brook. On a terrace looking out over a wide valley, this was among the earliest settled farmland in town.[13] Part of it is cleared land even now—a distinct rarity in Jamaica today.

Following the twists and turns of land acquisition in the town records, it is possible to trace how Peter Chase patched together the farms he would ultimately leave to his own sons. By 1843, he had begun the process of turning over control of the land to the next generation. His younger son Martin was now listed as the tenant of a 150-acre farm "commonly called the Smith lot," not far down the road.[14] At some point before 1850, Peter Chase's oldest son Daniel took over the home farm. The 1850 census recorded the role reversal in the family: Daniel was now the owner of the farm, and his elderly father lived with him.

Forest to Farm

The brothers Chase had succeeded in creating workable farms on paper, but it also took time to carve usable farmland from the woods. The 1790 Jamaica grand list included forty-six property-owning men: just ten of those men owned any cleared land at all.[15] Neither Elisha nor Stephen Chase had yet cleared a single acre. Three years later, they had cleared just four acres apiece.[16] After another fifteen years had passed, the 1806 grand list showed more progress: Elisha now owned twelve cleared acres, Stephen nine, and the youngest brother Peter six. Over the next several years they continued to add a few acres at a time, and by 1813—a full thirty years after they had made their first land purchase—the three brothers owned a total of forty-six acres of cleared land. By 1827, Peter had accumulated twenty cleared acres on his own home lot and purchased two other cleared lots of thirty-five and twenty-five acres from neighbors. After forty-four years of work on his and his brothers' land, he had something approaching a fully realized farm.

It was not that the brothers lacked experience. They must have been well prepared to clear land and make farms. They had almost certainly been part of a similar process while they were children back in Petersham. And as it happens, Petersham's historical agricultural transitions were carefully documented in the early twentieth century by researchers affiliated with the Harvard Forest, who created a series of well-known dioramas that depict changing patterns of land use over time.[17] One of the best-known of those dioramas illustrates the practices the Chases must have learned in Petersham and then transferred to Jamaica (see Figure 1). Perhaps around 1805, after the first stages of buying, selling, and clearing were all completed, the Chase brothers' farms might have looked a little like the one pictured in the Harvard diorama. The first rough plots of corn, wheat, and rye planted around the tree stumps had been replaced by tilled fields; cattle were grazing in the half-cleared woods; an orchard had been planted. Perhaps the brothers started off with log cabins, but as early as 1782 a sawmill was in operation near them, so their new houses may already have resembled the frame building pictured in the diorama.[18]

In the early stages of development, farms like these may have looked primitive, but they were already an important part of the regional economy. If the brothers Chase followed the typical pattern, they burned many of the trees they cut and

FIGURE 1. "An Early Settler Clears a Homestead—1733," Harvard Forest Diorama 2, Harvard Forest Archives. This is the second of twenty-three dioramas belonging to the Fisher Museum in Petersham, Massachusetts, created in the 1930s to illustrate the rise and fall of farming in central New England. This idealized image depicts Petersham around 1733, two generations before the Chase brothers left for Jamaica.

sold the ash. (One later account suggests that, rather than cutting and burning indiscriminately, early settlers were quite selective in the trees they cut for potash. "Good timber was just as valuable then as it was now, and the old-timers had better sense than to destroy it. What the first settlers burned was . . . either hollow or rotten or shaky or poor stuff."[19]) An important ingredient in a variety of industrial processes, potash found markets far beyond the forest settlements where it was produced. As early as 1794, Jamaica settlers could trade their potash to a merchant just over the border in Townshend, who would exchange it for the salt cod, tobacco, and rum he obtained from downriver. And as one early resident explained, potash was exclusively a cash crop: "The proceeds of these salts were a very important item on the credit side of our store bills."[20]

Like the families they came from in southern New England, northern farmers brought with them the expectation that they would operate partly within and partly outside the marketplace. They may have recalled a time when land had been allotted by an older, non-cash system, but in this generation, they bought it in a speculative market. They needed cash to acquire their land, and cash to pay taxes on it. If their families were to thrive, they would need to produce for market. But they also expected to provide for most of their household needs—not just in that early pioneering stage, but permanently, as part of the promised stability and security that would make it worthwhile to clear those farms in the first place.

What the Land Was Good For

An observer today might perceive little in the way of farming opportunity in the rugged land that became Jamaica, but a farmer skilled in the mixed agricultural practices of eighteenth-century New England would have recognized many possibilities there. Early published accounts were lavish in their praise of the farmlands of Vermont, and they explicitly included hills like these in their appraisals. One writer declared that the soil of upland Vermont was "rich, moist, warm and loamy." As soon as it was cleared of woods, the soil would grow all kinds of crops—and even after that first flush of fertility was over, "it naturally turns to rich pasture or mowing."[21] To be sure, describing Jamaica specifically, Zadock Thompson's 1824 gazetteer (definitive in its day) was a little more guarded in its assessment.[22] Thompson acknowledged that the "surface of the township is broken and mountainous, and the elevations rocky." But the soil itself, he concluded, was "in general, warm and productive."[23]

That description placed Jamaica somewhere in the middle ranks of the soils of Vermont as Thompson evaluated them. The land downstream from Jamaica was better, including "large and fertile meadows" in Westminster and a "large tract of excellent intervale land" in Putney. On the other side, directly west of Jamaica and higher still in the mountains, the town of Stratton's soil was worse—"cold and generally unprofitable."[24] In the next generation, Stratton would be described as so "extremely irregular and mountainous" that it was "in some places unfit for habitation." Jamaica's soil, in contrast, would still be called "warm and productive."[25]

Above all, this was grazing country. One writer predicted in 1808 that the "mountainous tracts" of southern Vermont would one day "be reckoned among the best of our grazing townships."[26] In Thompson's judgment, there was "perhaps no part of the world better adapted to the production and fattening of horses, cattle and sheep, than the hills and mountains of Vermont."[27] The cool summer temperatures and ample rainfall of the higher elevations would support rich grasslands.

In the first years of settlement, the farmers of Jamaica were not yet in a position to test that assertion. In 1790, the Chase brothers held only a single team of oxen between them. But by 1806, the three brothers were already being taxed for a total of twenty-five cattle. And by 1829, the youngest brother Peter's taxable property included his own substantial herd of twenty-one. That mattered, because cattle were essential to New England farming practices.[28] For the household, they filled a wide array of needs, providing milk, butter, cheese, and beef, along with the oxen that were necessary to the work of the farm. They could be precisely targeted to serve farm needs: heifer calves as replacement milk cows; bull calves as breeders, oxen, or meat to be consumed at home or traded to a neighbor. Even worn-out oxen and elderly milk cows were converted to beef.[29] To satisfy all those home and neighborhood needs, the average Jamaica farm in 1830 maintained around seven cattle, typically including a pair of oxen and several milking cows.[30]

But the importance of cattle extended far beyond subsistence needs and local exchanges. From almost the earliest days of settlement, southern New England farmers had sold cattle, beef, cowhides, and dairy products in local, regional, and international markets.[31] Northern New England was drawn into that larger rural economy in its first generation of settlement. By the early nineteenth century, Vermont farmers were sending large numbers of animals (mainly cattle, but also sheep and pigs) to the rapidly developing central meat market in Brighton outside Boston. As early as 1808, one author estimated that "from 12 to 15,000 head of beef cattle are driven every year from this state to the Boston market."[32] Zadock Thompson affirmed that Boston had "always" been the "principal mart for our beef, pork and mutton."[33] Gazetteers in both 1824 and 1839 reported that Windham County, where Jamaica is located, ranked second in the state in cattle production, behind Windsor County directly to its north (reflecting the importance of the Connecticut River valley as a route for cattle driving).[34] Some of Peter Chase's twenty-one cattle were doubtless destined to supply household needs, but some probably produced dairy products or beef for local sale, and others may have joined a drover's herd on the road to the Brighton market.[35]

But by the end of the 1820s, it was already becoming clear that Vermont's good grazing land would be adapted to a different kind of livestock enterprise, potentially more lucrative—and more transformative—than the Brighton beef market. As early as 1824, Thompson had predicted the advent of a new commercial venture: "should government see fit to take our infant manufactures under its fostering care"—that is, by passing a protective tariff—"the raising of wool will probably, at no distant period, constitute a principle branch of agricultural employment."[36] In 1829, following that same hunch, Peter Chase grazed on his recently-created pastures not only those twenty-one cattle, but nineteen sheep—a small number in comparison with what was to come, but enough to put him in the vanguard of Jamaica farmers that year.[37]

Wool Booms and Busts

From the beginning, hill farmers like those in Jamaica had been involved with local and regional markets; but raising sheep would introduce an altogether different level of market participation. Enthusiasm for wool production began as a short-lived boom generated by the embargoes on British products leading up to the War of 1812, but that initial bubble burst after the war ended, as American markets opened again to British wool. A second and more significant expansion began shortly thereafter, when Congress passed the first of several increasingly strong protective tariffs to support the new industry. New England farmers, and especially Vermont farmers, were quickly drawn into the enterprise, now coming to specialize in the imported Merino stock that produced high-priced luxury wool. In 1824, when he predicted this second boom, Zadock Thompson estimated that there were already about 475,000 sheep in Vermont. By 1836, the state counted over a million, and by 1840 the number was reported to be 1,681,000.[38] In those years, Vermont became briefly famous for its crowded sheep pastures.

But as transformational as the business was, sheep did not have an equally great impact everywhere. Agricultural historians have sometimes assumed that because sheep were naturally suited to grazing the rocky highlands, that is where they were mostly located.[39] In fact, the statistics reveal almost the opposite pattern.[40] Wool was a high-risk enterprise. Especially in the early years, it seemed to promise dazzling profits—but not to all farmers. Success was more likely to reward the efforts of well-established, prosperous farms with access to start-up cash and connections.[41] To become a breeder of one's own stock required even more money and expertise. First-class breeding rams at the height of the boom could cost thousands of dollars. The best most farmers could do would be to breed their own sheep with the Merino or partly Merino rams of wealthy neighbors and hope to improve their flock gradually.[42]

Where sheep farming did "boom," the business had striking social consequences. It led to the consolidation of farms, as expanding sheep flocks required more acreage and fewer workers than before. Prosperous farmers bought out their neighbors, who left for New York, Illinois, or Lowell.[43] Wages stagnated while real estate prices went up, making the path upward from laborer to proprietor a more difficult one. Profits accrued disproportionately to the wealthiest farmers. The difference in social status between farm laborers and farm owners became more pronounced. While successful sheep farmers joined a new middle class of rising expectations and new opportunities (and even embraced new forms of religious expression), their neighbors often found themselves shut out. Few of the signs of social disruption that characterized the sheep boom areas appeared in Jamaica.[44]

The greatest investments and the largest flocks in Vermont were to be found in the rich lowlands of the Champlain Valley, with a second center in the floodplains of the Connecticut River, around Weathersfield, where the prominent merchant and investor William Jarvis first introduced imported Merinos in 1811. The 1836 statewide count found that the largest number of sheep were crowded onto the broad fields of the town of Shoreham, in the heart of the Champlain Valley, on the richest and smoothest soil in Vermont: 26,584 sheep in one town.[45] Jamaica's farms, in contrast, were reported to contain just 3,863 sheep in the 1836 register. That is not an insignificant number, to be sure.[46] But it amounts to just two and a half sheep per capita, while Shoreham's per capita figure that same year was nearly thirteen. As for the social consequences of that difference: sheep-mad Shoreham's human population dropped over 20 percent in the single decade between 1830 and 1840. In those same years, Jamaica's population held stable, even growing a little.[47]

Still, there were clearly some farmers in town who were willing to take a risk on the new enterprise. From the mid-1820s through the 1830s, Jamaica grand lists consistently reported over two thousand sheep in town—an average of around eleven per household (see Table 1). Of course, they were not evenly distributed. A few farmers held large flocks of eighty or a hundred, while most held few or none.[48] In 1837, a severe depression brought the first serious check to the sheep boom, causing a temporary decline in wool prices, but the numbers of sheep in Jamaica did not appear to respond much to that downturn; a slight dip in 1839 was followed by renewed growth.

Table 1. Sheep in Jamaica, 1830–1880

GRAND LISTS	# SHEEP IN JAMAICA
1830	2351
1831	2512
1832	2141
1833	2289
1834	2512
1835	3055
1836	2696
1837	2757
1838	2790
1839	2330
1840	3274
1841	3360
1842	3255
1843	3975
1844 no data	--
1845	4318
CENSUS	
1850	2258
1860	2464
1870	1515
1880	1646

In fact, Jamaica's flocks continued to grow for several more years, reaching an average of nearly fourteen per household in 1845. That continued growth was a bit anomalous. In Vermont as a whole, the number of sheep had already begun to decline several years earlier, in late 1840. With both political and economic threats on the horizon, the biggest—perhaps the best informed—sheep investors in the state were evidently responding to warning signs that might not yet have reached Jamaica sheep holders. (One can imagine sheep farmers following the political fallout after the 1840 presidential election with particular care, as the victory for Whig pro-tariff forces seemed to evaporate with the sudden death of the new president.) Sharply lower prices for wool in 1839 and 1840 were followed by a series of disastrous cuts to the protective tariffs for wool. Between 1841 and 1846, tariff supports for the industry were lowered and finally completely removed, and prices of wool collapsed.

In the areas of the state most deeply invested in sheep, the effects of these combined events were sudden, dramatic, and permanent. Within a few years after the tariffs were removed, sheep numbers in the town of Shoreham had dropped by 50 percent. Tales abound of sheep being sent to the slaughter by the thousands in a single season.[49] And the declines continued: statewide, sheep numbers in the state dropped by nearly 40 percent between 1840 and 1850, and by another quarter between 1850 and 1860.[50] Jamaica followed a somewhat different pattern. Numbers of sheep in Jamaica did decline, to be sure: from their high point in 1845, they dropped by almost half in 1850 (see Table 2). After that, however, Jamaica's pattern began to diverge from the statewide declines. Sheep numbers in Jamaica rebounded slightly between 1850 and 1860; they dropped by another third between 1860 and 1870 (after a short-lived but intense price spike driven by Civil War demands for wool); and then rebounded again between 1870 and 1880. In all these years, statewide numbers continued their consistent downward slide. Overall, from "boom" to "bust," it appears that Jamaica farmers were responding comparatively slowly—perhaps at times, not at all—to the signals they were receiving from the wool markets (or from Congress).[51]

Table 2. Comparative Decline in Sheep Holdings, Vermont and Jamaica

	VERMONT	JAMAICA
1840 to 1850	-40%	-30%
1850 to 1860	-26%	+8%
1860 to 1870	-30%	-36%
1870 to 1880	-27%	+14%
TOTAL % CHANGE 1840 TO 1880	-74%	-50%

There were always a handful of Jamaica farmers who wholeheartedly embraced the wool market, and presumably its new path to commercialization. But they were not the norm. From the beginning, the numbers of sheep in town were quite unevenly distributed. A few farmers held large flocks of a hundred or more, while most held just a few sheep, or none. At the height of Jamaica's "boom" in the 1840s, about half of households were assessed a tax for sheep.[52] After the "bust," ironically, sheep were actually more widely distributed in town—65 percent of farms in 1850 and 60 percent in 1860 held some sheep (see Table 3). But that broad distribution declined after midcentury. All in all, the pattern is clear: Stephen Hescock, who held ninety-six sheep in 1827; Calvin Briggs, who held 211 in 1846; Pliny White, who held 114 in 1850; and Joseph Covey, who held 140 in 1870 were operating on a scale much larger than that of most of their neighbors. For those few, heavy investment in sheep might well have ushered in a new commercial reality of boom-bust cycles like those experienced in the prosperous lowlands. But the great majority of households in Jamaica appear to have maintained a more conservative approach. Selling a few lambs or fleeces from a small flock might bring in a little money in a good year, but an investment in five or ten sheep was unlikely to bring the farm to ruin when another price crash occurred.

Table 3. Distribution of Sheep in Jamaica by Size of Flock per Farm

	0 SHEEP	1–10 SHEEP	11–49 SHEEP	50 OR MORE SHEEP
1850	35% of farms	37%	25%	3% (7 farms)
1860	40%	40%	12%	8% (17 farms)
1870	53%	30%	14%	2% (5 farms)
1880	62%	24%	12%	2% (6 farms)

The Landscape of Sheep

In many parts of New England, the adoption of wool as a cash crop was associated not only with commercializing agriculture, but also with widespread deforestation. The third Harvard diorama was intended to illustrate what that might have looked like, depicting central Massachusetts in 1830 as a wide-open treeless landscape with long sight lines of distant hills (see Figure 2). Perhaps some parts of Jamaica's landscape looked something like that twenty or thirty years later. On David Robbins's 225-acre farm, for example, there were 250 sheep in 1860.[53] The agricultural census recorded that Robbins sheared twelve hundred pounds of wool that year, by far the largest harvest in Jamaica, over three times more than the next largest producer in town, and a very high wool yield per sheep for the time. No surprise, then, that Robbins reported to the census that his land was 90 percent cleared. At about 1.25 sheep per acre, Robbins's farm was not necessarily overstocked by the standards of the day, but he was certainly using the land intensively. His land, at any rate, must have looked something like the third Harvard diorama.[54]

FIGURE 2. "Height of Cultivation for Farm Crops—1830," Harvard Forest Diorama 3, Harvard Forest Archives. The third of the twenty-three Harvard Forest dioramas depicts the fully deforested landscape at the height of the boom in sheep raising. Curiously, the image shows none of the animals that presumably created this landscape. Perhaps that is because Petersham did not actually specialize in sheep. It did not reach its actual high point of 70 percent deforestation until fifty years later, in 1880, and its clearance was caused by cattle grazing, not sheep.[55]

Still, it is difficult to imagine that the town's comparatively small total numbers of sheep would have been enough to create a completely open townwide landscape like the scene that is depicted in the diorama. And in any case, Jamaica's open hillsides were not being grazed exclusively by sheep. In other, more sheep-heavy regions, at the height of the boom of the 1830s, it was not just the human population that declined dramatically, but also the number of cattle, as farmers filled up their pastures with sheep instead.[56] By one count, in the years between 1832 and 1837, cattle numbers declined by 54,000 in Vermont.[57] In Jamaica, in contrast, rather than declining, cattle numbers grew by a third between 1830 and 1845, suggesting that Jamaica farmers did not give up cattle to make room for sheep, but rather added sheep to their fields alongside their growing herds of cattle (see Table 4).

It was common for farmers with substantial sheep flocks to own many cattle as well. An early example was Peter Chase, who reached his largest holdings of both cattle and sheep in 1832, with thirty-three cattle and twenty-four sheep. By 1850, similarly, Squire Gleason grazed a herd of forty-two cattle alongside his forty-five sheep; his four hundred acres were two-thirds cleared at that point. And in 1860, Justus Holden held fifty-six cattle and one hundred forty sheep on his very large plot of land. (He owned 640 acres, less than half cleared, in an area west of the central village where several other large sheep holders lived.)[58] Collectively, in midcentury, Jamaica farmers held almost the same number of sheep as cattle: between ten and twelve cattle per household, and around eleven sheep (see Table 4). Given the different sizes and grazing needs of the two animals, it is possible that in some parts of town, cattle may have had a much greater impact on Jamaica fields than did sheep. Although sheep were thought to be particularly well suited to the stony pastures of the highlands, the owners of those pastures nevertheless most often maintained a pattern of mixed animal husbandry, perhaps pursuing a more diversified production system that would allow them to respond more flexibly to market risks.

Table 4. Cattle and Sheep Holdings in Jamaica, Compared with Rates of Deforestation and Reforestation

	# SHEEP	SHEEP PER FARM	# CATTLE	CATTLE PER FARM	% FARMLAND "IMPROVED"
1850	2258	11	2632	12.6	59%
1860	2437	10.9	2406	10.7	76%
1870	1515	7	1810	8.4	70%
1880	1646	7.8	2212	10.5	61%

At some point around 1860, Jamaica reached its high point of land clearance. Between 1850 and 1860, the percentage of "improved" farmland in town shot up from 59 percent to 76 percent. ("Improved" land is not an exact equivalent to cleared land, but in the northeast, at least, it can offer a close approximation.[59]) A town that retained only a quarter of its land in woods was probably pushing the limits of its soil's resilience and ability to hold moisture—particularly in a place like Jamaica, with its stony uplands and steep slopes. (Indeed, George Perkins Marsh's groundbreaking warning of that threat, *Man and Nature*, was issued in 1864, at nearly the exact high point of clearance in Jamaica.[60]) As with the distribution of sheep and cattle, however, the degree of clearance varied a great deal from farm to farm. Seventeen properties (all farms of under fifty acres), held no wooded land at all in 1860. But fifteen farmers held a hundred acres or more in wooded land, and many of these holdings were quite large (ranging from 140 acres to 640). And there were also large parcels of land—often rugged or remote—that were not intended as present or future farms, but as timber lands, held for current or future logging. At least some of that land was not included in the agricultural census, and it had a development timeline of its own.

In midcentury, then, Jamaica must have presented a much more varied appearance than the landscape depicted in the third diorama. Parts of town were indeed open expanses of treeless pastures with grazing flocks of sheep, herds of cattle, or both; but in other areas, high ground was still covered in wide swathes of woodland. Everywhere else, small farms created a patchwork landscape: a few acres of tilled land; grazing land and meadows; a woodlot that included the maples for the farm's sugar crop. By 1870, moreover, the tide had already begun to turn, although at first only gradually. By 1880, improved land was down to 61 percent of farmland. The average wooded acreage on Jamaica farms had gone from thirty-three acres to over fifty acres in those twenty years. (In comparison, statewide records showed very little change in cleared land over this period.)

There is certainly a relationship between the rise and fall of sheep and the return of trees to farms, but it is not a simple one. Between 1860 and 1870, sheep numbers for the first time declined dramatically in Jamaica, and that change coincided with the end of the increase in cleared land. It also coincided with an almost equally sharp downturn in the number of cattle (see Table 4). But in the next decade, both sheep and cattle numbers rebounded—and the reforestation nevertheless continued. Whether because of sheep, cattle, or something else, Jamaica's land use seems to have shifted relatively abruptly from rapid clearance to the early stages of reforestation. By 1880, observers were already beginning to express concerns about the rapid reforestation of the hill farms in southeastern Vermont. Which was better for the land and its people—deforested or reforesting—would be the subject of much debate in the years to come.

Transition to Dairy?

Traditionally, agricultural historians have described the late nineteenth century as a time when Vermont farmers (and to some degree New England farmers generally) made a full-scale transition from commercial sheep production to commercial dairy operations—first specializing in cheese, later in butter and cream, and finally in fluid milk, as centralization, railroads and refrigeration gradually made it easier for competitors in other regions to undercut prices for each of those products. From that standpoint, one might see Jamaica's midcentury mix of sheep flocks and cattle herds as a movement toward a future dairy specialty. But that is not really what happened. Speaking in terms of a "transition to dairy" is more or less accurate on the statewide level, but the term creates a false impression that most farms had become fully commercialized at midcentury. The narrative of the "transition to dairy"—like that of the statewide sheep boom—obscures the widespread continuity of farms that were neither fully "sheep" nor fully "cow." That kind of farm combined diversified, mixed animal husbandry with a casual and relatively cautious engagement with a variety of local and more distant markets; in Jamaica, it was the most common way to farm.

In contrast, one Vermont town that did make the classic transition from commercial sheep to commercial dairy was Chelsea, the subject of Hal Barron's groundbreaking 1984 study, *Those Who Stayed Behind*. Chelsea is an example of how difficult that transition could be, even in a town that was considerably more committed to commercial farming than was Jamaica. Chelsea was comparable to Jamaica in size and history, but it enjoyed a somewhat more favored location in the valley of the First Branch of the White River, a broader valley and somewhat gentler landscape than Jamaica's West River valley. Indeed, Zadock Thompson had used the same phrase—"warm and productive"—to describe both towns' soils. But Chelsea was far more committed to sheep than was Jamaica.[61] Even in 1880, after decades of declining prices for wool, Chelsea farmers still averaged twenty-five sheep per farm: Jamaica farmers averaged under eight. Two-thirds of Chelsea farmers still raised some sheep in 1880: just 38 percent of Jamaica farmers did so.[62] Chelsea farmers held onto sheep as a primary business until one last steep drop in the price of wool near the end of the century finally forced their hand. In the end, however reluctantly, Chelsea farmers did make the difficult shift to commercial dairy farming, and Chelsea became a classic twentieth-century Vermont farm town, fields dotted with black-and-white Holsteins. Since the early nineteenth century, Jamaica and Chelsea farmers had taken different paths; now those paths would diverge even more from one another.

A few Jamaica farmers did become intensive dairy producers in the second half of the century, and their story suggests that it was not impossible. Arad and Fannie Wood, the third-generation owners of the Chase farm, presided over such a farm. By the 1880 census, their farm counted ten milking cows, placing it well above the statewide average of six per farm. In 1860, the household made two hundred pounds of butter; in 1870, five hundred pounds. By 1880, the family had dramatically increased production, making twelve hundred pounds of butter and three hundred pounds of cheese. (Overall, Vermont farms produced an average of just 710 pounds of butter on their farms that year.) At that level of production, Fannie Chase Wood was essentially presiding over a small-scale creamery of her own; indeed, the household now included a hired couple who lived on the farm, likely to help with the extra work of butter making.

But this was not the direction most farmers in Jamaica were taking. Of course, almost all farm households milked cows. In 1860, all but six Jamaica farms (out of 223) kept at least one milk cow.[63] Nearly everybody made some butter.[64] But the average production of butter on Jamaica farms in 1880 was far lower than the statewide average.[65] And it was no wonder that Jamaica farmers produced less than the state average: they kept only half the statewide average number of milk cows (just over three per farm as compared to the statewide average of six). Clearly, Jamaica farmers were not moving in any concerted way toward commercial dairy production.

Dairy farming required one resource that was very hard to come by in the late nineteenth century: extra labor. As Barron demonstrated in his analysis of Chelsea, farmers there hesitated to switch to dairy in large part because of the high labor demands associated with modern dairy operations, at a time when farm labor was increasingly scarce and expensive.[66] Over time, the choice to milk cows for commercial production would draw farmers into increasingly complex and expensive innovations: to improve their feeding and their stock; to keep milking through the winter (requiring a larger and more stable winter feed supply); to raise the butterfat level of their milk (and ultimately to lower it in a couple of decades as they transitioned to fluid milk production); and to purchase off-farm fertilizers, feeds, and machinery, including expensive investments in silos and ensilage equipment.[67] Jamaica farmers were even less likely than their Chelsea counterparts to find it possible or desirable to shift to such a system. They had never gone all in on sheep (collectively at least), and they were not making an orderly transition to commercial dairy, either.

What They *Were* Doing

At first glance, the information available from nineteenth-century agricultural censuses seems to suggest that Jamaica farmers were not putting a great deal of effort into any cash crops. In 1880, when the most detailed census information was collected, Jamaica's town-wide average yields of almost every crop were lower than state averages.[68] For some important field crops the pattern was strikingly clear. Notably, Jamaica farms averaged relatively low yields of the ubiquitous potato, a critical staple of household self-sufficiency and one of the most reliable of all crops for surplus sale. Almost all Jamaica farmers grew some potatoes, but the town's overall yield per acre was less than two-thirds of the state average.[69] On top of that, the yields were wildly variable: potatoes could yield anywhere from ten to two hundred fifty bushels per acre—from a mere 8 percent of the state average to well over 200 percent. The same pattern can be seen in corn, a crop that was close to a necessity for both human and animal consumption.[70]

One obvious way to explain these relatively low yields would be to blame the cold weather, steep hillsides, and stony soils of the highlands. Certainly, the geographers, soil scientists, agricultural economists, and geologists who would descend on the hill farms in the early twentieth century would have much to say on these matters (a subject for chapter 2). But land and climate may not be the best (or at least not the only) explanations for the low and erratic yields of field crops. Like most nineteenth-century farmers, Jamaicans were forced to choose carefully which fields they would manure, and how frequently.[71] (Indeed, the success of New England farming had always depended on the careful distribution of animal manure, whether gathered in the barn to be spread on the tilled fields, or scattered by grazing animals in the distant pastures.[72]) That might help to explain the widely diverging results on different farms, where a decision might have been made to stint the potatoes and feed the corn, or vice versa.

There may have been other resources, too, that were in even shorter supply than manure. The production levels of maple sugar and other woodland products cannot be said to depend directly on the fertility of the soil as do corn and potatoes. Nearly all of Jamaica's farms held at least a few acres of woods; even in 1860, at the height of clearance, the average per farm was still thirty-three acres of woodland. By 1880 it was fifty.[73] Yet Jamaica farmers not only reported lower field crop yields, but they also cut significantly less cordwood than the state average.[74] Even the town's maple sugar production (a crop that would become a mainstay in the twentieth century) was a little low in comparison to state production.[75]

There may have been other reasons, but it seems likely that most Jamaicans simply cut less wood and tapped fewer trees than they had available.[76] A 1915 count reported, for example, that Jamaica farms tapped only a little under two-thirds of the maple trees available to them that year.[77] And another later account suggests that maple output was not limited by the productivity of the trees, but by access to labor. Two hundred to four hundred pounds of sugar was the normal production of "ordinary small farms" in southeastern Vermont at the turn of the century because that production level required setting out and caring for between two and four hundred buckets—about the limit of what one man could do without hiring extra labor.[78] The census of 1860 was the first year to record a population decline in Jamaica, but by 1880 the town had already lost 22 percent of its people, and with it, much of the casual labor pool that might have assisted farm households with sugaring.[79] Taken together, these low numbers seem to suggest that Jamaica farmers had to make carefully calculated decisions not only about manure, but also about how much labor to expend and on what crops.[80]

Drovers and Dealers

There was only one category in which Jamaica farmers collectively reported a figure higher than the state average. That was the category that bore the name "other cattle." In the census, "other cattle" denoted all the cattle that were not currently working as oxen or milking cows—a catch-all category that included young animals being raised to replace milk cows or oxen, bulls for breeding, beef animals, and animals past their prime. While Jamaica's farms averaged only three "milch" cows per farm in 1880 (half the statewide average), they reported an average of a little over six "other" cattle (25 percent higher than the statewide average). These are not mutually exclusive categories, of course, particularly since many Vermont dairy farmers raised their own replacement stock and often raised replacement stock for others as well. But the figures are clearly different from those in other parts of Vermont. As the state as a whole was moving gradually toward more milk cows for dairy production, Jamaica farmers actually held slightly fewer milk cattle in 1880 than they had in 1860 and 1850.

This is a statistic that may help to make sense of the others. From the earliest days of settlement, through the sheep booms and busts, and into the twentieth century, cattle played a vital and sustaining role on Jamaica farms. By 1880, four-fifths of Jamaica farmers held at least some animals in the "other cattle" category, ranging in numbers from 1 to 107. To be sure, just as most Jamaicans held only small flocks of sheep, most also raised only small herds of cattle. In 1880, only about a fifth of the farmers who held "other cattle" owned herds of eleven or more, and only twelve individuals held twenty or more. Half of Jamaica farmers held five or fewer "other cattle"—just about the number required to supply their own household needs for meat, oxen, and future milk cows. Another quarter owned between six and ten "other cattle," perhaps intending to increase their dairy herd to sell the surplus cream or butter, or to sell beef or livestock to a neighbor or a dealer. Just as with a flock of five or ten sheep, a few cattle could produce (or become) market products without demanding a long-term plan or a big investment of cash. A farmer operating this way could shift resources back and forth based on home needs, current prices, the quality of the hay crop, or the offer of a dealer passing by.

For such an operation, there would be less reason to modernize the animals or the production process than there would be for someone hoping to become a dairy specialist, and less pressure to breed more productive stock. In such operations, calves might be left with their mothers (rather than separated at a young age so the farmer could have the milk); the priority of the farmer might not be to maximize milk production but to fetch a good price for a well-fed young animal. Farmers with only a few milk cows probably fed them, along with other stock, almost exclusively on grass and hay, often grazing them in woods or in partly re-grown fields—not the current "best practice" in 1880 for high-quality cream production, but fine for mixed operations like these.[81]

The owners of the largest cattle herds were operating a different type of business, catering to a market that had been known to their fathers before them. In midcentury, the chief cattle market was still downhill in eastern Massachusetts. In 1849, *Scientific American* reported that the Vermont Central Railroad had now made it possible for beef growers in Vermont to move their fattened cattle from the railroad depot at Bellows Falls to Boston without "the loss of any of their flesh" along the way, noting that the arrival of a train of forty-four beef cars had arrived in Boston just the week before.[82] As early as 1851, a regional newspaper noted that "Vermont beef cattle, transported in special cattle cars, led the rest of New England in the Brighton cattle market."[83] Jamaica was not directly accessible by railroad until 1879, when the West River Railroad began operating between Brattleboro and South Londonderry, but its cattle could travel by road to connect with cattle trains in Brattleboro or Bellows Falls.

Indeed, in a fanciful passage in his 1855 book *Walden*, none other than Henry David Thoreau described in almost mystical language the connection between northern hill farmers, railroads, and the cattle trade. The Vermont Central cattle train, Thoreau wrote, carried not just the "cattle of a thousand hills, sheepcots, stables, and cow-yards," but also the essence of the north country itself, "as if a pastoral valley were going by." The train brought south "all but the mountain pastures, whirled along like leaves blown from the mountains by the September gales," and even the sound of the drovers' dogs all the way back home, "barking behind the Peterboro' Hills, or panting up the western slope of the Green Mountains."[84]

By the time the railroad made its way to Jamaica in 1879, however, it had also linked midwestern and even far western farmers to eastern markets, undercutting regional prices of beef and diminishing the northern beef trade with the southern New England cities. In Thoreau's melancholy assessment of the impact of the railroad, "So is your pastoral life whirled past and away." But in Jamaica, the "other cattle" were likely still traveling south, though some were now destined for different markets. The 1884 Child's *Gazetteer* named fourteen Jamaica farmers who specialized in three different aspects of the cattle trade. Five of the farmers were described as breeders of Durhams, Jerseys, or "Dutch" cattle, modern stock for commercial dairy production.[85] Three others dealt in "young stock," selling replacement cows to dairy farms. (Vermont dairy farms typically raised their own replacement stock, and frequently also sold to dairy farms in southern New England.[86]) The remaining six men were described as "cattle dealers": those who conveyed their neighbors' cattle to markets of all kinds.

These were the handful of farmers who reported those very large numbers of "other" cattle to the 1880 census taker. Some appear at first glance to be dairy farmers. Nelson Starr, for example, owned thirty-two milk cows in 1880, the largest owner of dairy cows in town. But Starr sent only a very minimal one hundred gallons of milk to the creamery, nor did his household make butter at home. In addition to his thirty-two milking cows, he also held thirty-five "other" cattle (and two oxen). Starr might be called by that very old term "grazier"—a person who rears or fattens animals for market. By the late nineteenth century, those farmers were usually called "cattle dealers." Their neighbors relied on them to market the small numbers of animals they raised for market as beef or young stock.

The largest herd of "other" cattle in town belonged to Sullivan Foskett. He counted 107 animals, but there were many more cattle coming and going on his farm: Foskett reported in 1880 that he had both bought and sold as many as three hundred that year. His household actually did make prodigious amounts of butter and cheese from the cream produced by his own fifteen dairy cows. But the scale of his enterprise was due mostly to his work as a "dealer in livestock," as he was described in the 1884 Child's *Gazetteer*. Like other farmers of this type, Sullivan Foskett dealt in both cattle and sheep.

A similar story is that of another large-scale stock owner, Stillman Clark, who raised 150 sheep and sheared 400 pounds of wool in 1880. But Clark was described by the 1884 Child's *Gazetteer*, not as a "wool grower," but a "cattle dealer." Clark's death certificate used an older word for his occupation: it named him a "drover"—one who drives animals to market on the hoof. Perhaps there was a reason why such an old-fashioned term seemed appropriate for farms like his: they were big operations, but they required little in the way of modern inputs, capital, or hired labor. In Thoreau's description of the cattle and sheep being sent from the northern mountains, he suggested that the actual human drovers of past days had all lost their purpose now that the railroads had arrived: "on a level with their droves now, their vocation gone, but still clinging to their useless sticks as their badge of office."[87] But being a "drover"—now called "cattle dealer"—was still a viable trade in late nineteenth-century Jamaica. Decades later, one of Stillman Clark's descendants would use the term "cattle trader" for his own occupation. Even as late as 1950, when only sixteen remaining Jamaica residents described themselves as farmers, four of those men still used the term "cattle trader" to describe their operations.

Complex Systems

In Jamaica, raising cattle—whether grazing, driving, dealing, or even milking—was the most constant feature of farming in the nineteenth century. Built on that practice was a diversified and complex farming system, one that allowed for the existence of highly individualized strategies. Indeed, it may be the sheer complexity of this malleable system that ultimately explains Jamaica's low (and erratic) crop yields. To take one example: the brothers Aurelius and Osmore Howe kept their large flocks and herds on 175 acres high up on a rocky hillside in East Jamaica, the earliest commercial center to emerge in the town back at the beginning of settlement. (The Howe clan had deep roots in that neighborhood, and the brothers were surrounded by neighbors and relatives who also kept sheep. No doubt that was part of the appeal of the business: the Howe brothers operated one of the few farms for which the records show a consistent commitment to sheep raising over time.[88]) The two brothers were hardly what would ordinarily be considered subsistence farmers: the value of their livestock was almost double the town average, and the value of their farm was more than twice that of their neighbors. Yet they could not really be regarded as commercial specialists either, certainly not in dairy, and not even in sheep. That is presumably why the 1884 Child's *Gazetteer* identified them as "cattle dealers," gathering the neighbors' sheep and cattle for sale in some downhill market.

Their cash-producing flocks and herds required substantial grazing land. The brothers needed large quantities of hay for winter, and they grew rye, oats, and corn for supplemental feed. But the rest of their farm operated on a smaller scale, mostly meeting their household's needs. They cut twelve cords of wood—in 1880 the average in Jamaica, and probably close to the minimum required for heating the house and fueling the next year's maple operation. Osmore's wife Viorna gathered forty dozen eggs from their henhouse and made 350 pounds of butter; they may have sold some eggs and butter, but probably most went to the household. Clearly, they intended some of the surplus of their other crops to go to local markets: they raised thirty bushels of apples, thirty bushels of potatoes, and two hundred pounds of maple sugar. (Nineteenth-century Vermonters did eat a lot of potatoes, apples, and maple sugar—but not as much as the Howes harvested that year.[89])

In none of the measurable categories did the Howe brothers surpass the relatively modest per-acre production averages of other Jamaica farms.[90] In some cases they did not even come close. But the Howes were nevertheless successful farmers: their production was clearly tailored closely to their needs. Their yields per acre of rye, oats, and corn were not impressive, but if one crop came up short, the system they operated generally ensured that their cattle would be well nourished with some other crop. It seems likely that the Howes measured their success, not by that absolute standard of "yield per acre"–which was after all a measurement only introduced by the agricultural census takers for the first time in 1880—but by their success in providing for the family table and for the farm's cash-producing flocks and herds.

On the western side of town, Joshua Stark operated within the same framework. His was a much more modest farm, worth only $400 (in contrast to the Howe brothers' $3,000 farm). Stark was sixty-eight years old, with just one sixteen-year-old son still at home to help, so he was not rich in labor, either. In contrast to the Howe brothers, who estimated that they had produced about $500 worth of goods "sold or used" that year, Stark had produced only about half that much. Clearly Mr. Stark was no "cattle dealer." But the Starks seemed to do well that year, too. They raised well over twice as many potatoes as the Howes, almost certainly bound for local markets. They (Abigail Stark, that is) made 650 pounds of butter at home—not quite up to the statewide average, but well above the average in Jamaica and much more than Viorna Howe did.

Butter was probably an important cash crop on the Stark farm. They owned two oxen, two milk cows, and seven "other" cattle—a standard herd size since the early days of the nineteenth century. From that herd, they were able to sell five calves that year, probably providing a good side income. To feed the animals, they raised a good crop of oats, an average crop of corn, and an excellent crop of buckwheat—by far the largest in town. Not many Jamaica farmers raised buckwheat, but it is a crop that can be used either green or dry for animal food, and one that performs well on poorly fertilized upland soils.[91] They sheared just two sheep that year, but those two fleeces weighed a whopping nineteen pounds.

Joshua Stark was not rich, but by several measures his farm seems to have been as productive as that of his wealthier neighbors. Like the Howe brothers, the Starks produced plenty for themselves and the animals they sold. And the Starks possessed an array of skills as complex as those of their more prosperous neighbors: they could even claim to outdo some farmers who operated on a larger scale. (Their buckwheat yield per acre was the second highest in town.) Their farm surely had many limitations—as their reliance on buckwheat hints—but their skill and knowledge enabled them to stretch those limits to fit their needs.

Whether the families' involvement with markets was little or great, local or distant, a daunting array of skills was required to keep farms like theirs operating.[92] If local markets changed, the Howes could move from sheep to cattle and back again without much trouble. On a smaller scale, Abigail Stark could increase her production of butter, perhaps hire a helper, or Joshua Stark and his son could tap more maple trees. While the modern term "resiliency" did not exist at the time, it seems to suit the situation.

This farming system was adaptable and durable, if only modestly productive in comparison to the farms on the flat lands of the Champlain Valley or the new land in the west. It was open to opportunities for windfall profits, but its practitioners were generally slow to commit to markets that were volatile, or that required large investments in equipment or off-farm inputs, or too much household labor. It provided room for individual farmers to experiment with a new specialty or to take a gamble (perhaps to try buckwheat rather than the traditional rye crop on poor land). Most of the time, though, such experiments operated alongside a generally reliable set of practices. This was a safety-first system, intended to feed and shelter the family—and pay the town taxes, and secure the futures of the children—with a minimum risk. It was a farming system built to provide stability when other aspects of life proved unstable, as they so often did.

The Landscape of Loss

The Civil War does not often play much of a role in northern agricultural and environmental historical accounts. But the evolution of Jamaica's agricultural system was interrupted by the disruptions of that war. The impact of the war is exemplified by the story of a family that was introduced at the beginning of this chapter: the family of the three Chase brothers—Elisha, Stephen, and Peter—who were among the first New England people to settle in Jamaica. Peter Chase, the youngest of the three brothers, passed his farm on to his son Daniel. By the 1850s, when Daniel and his wife took over the farm, families were not as large as they had been in the founding generation. Still, Daniel and Mary Jane Chase had three sons and a daughter—certainly enough, under more ordinary circumstances, to ensure the family's continuity on the farm. But the oldest son, Ossian, went to California, where he was working as a miner when the Civil War began. Ossian enlisted in the First California Cavalry regiment; he died in New Mexico in 1863. The youngest brother, George, enlisted shortly after his brother's death, only to die himself just a few weeks later, before he had left Vermont. By the time the war ended, fully one-tenth of Jamaica's total population had volunteered to fight for the Union. Obviously, the human toll was enormous, and the town went so heavily into debt to support the bounties promised to recruits that it took years to dig out of it.[93]

But the war was not the only problem. Daniel Chase's remaining son, his father's namesake, did not enlist, but he did not stay home to inherit the farm, either. Like so many sons and daughters of Vermont's hill farms in these years, the younger Daniel left his farm home for the West, traveling out to California in his oldest brother's footsteps.[94] Daniel and Mary Jane Chase were thus left with no sons to take over their farm. Shortly after the end of the war, they retired to a village house in nearby Wardsboro and deeded the farm to the husband of their recently married daughter, Fannie. Perhaps Arad Wood, the new son-in-law, felt at home with this stricken family; he himself had only recently returned from nearly four long years of war. Wood had enlisted as a soldier in 1861, traveling to the front along with his first wife Lydia Johnson Wood, who served as a nurse—until she contracted typhoid and died that same year. Arad Wood himself chose to remain in the army till the end of the war, returning home to Jamaica only in July 1865. He then married Fannie Chase and took over her father's farm, all within three months of his return.

The story of New England's Civil War losses is well known. Especially in small towns like Jamaica, the costs of victory were high, and the struggle defined the region for generations to come. But like other Vermont communities, Jamaica was deeply proud of its role in the war. The experiences shared by Arad and Fannie Chase Wood united them with the community. Through the nineteenth century and into the early twentieth, yearly celebrations marked the town's continuing united support for the sacred cause and all that had been done in its defense. As late as 1895, Arad and Fannie Wood held leadership roles in the Grand Army of the Republic veterans' organization and its ladies' auxiliary.[95] The memory of that common cause continued to strengthen community bonds for many of the residents of Jamaica for decades to come.

The Arad Wood Place

Toward the end of the century, as Arad and Fannie Wood's own children reached adulthood, Jamaica was clearly changing. In some ways, things seemed to be looking up. By the beginning of the new century, downtown Jamaica village boasted not only a railroad depot but an electric light company, a telephone company, a bank, and many new businesses. The Brattleboro and Whitehall Railroad had finally made it to town in 1879 after years of lobbying and fundraising, and the railroad had fostered a growing lumber trade. For decades, the railroad would deliver Jamaica wood to downhill businesses (until the lumber supply began to run dry, and downhill markets shifted to larger suppliers). In town, woodworking enterprises thrived. The 1884 gazetteer listed dozens of such operations, including shingles, axe helves, rakes, trays, chairs, washtubs and more.[96]

But as village businesses grew, the overall population of the town was falling rapidly. Between 1880 and 1900, over a third of the inhabitants of Jamaica left town. Farms, too, declined in number: more than 20 percent were lost between 1860 and 1900. The town's landscape must still have presented a patchwork of cleared and forested land, but the patches were changing places. As the large plots of timber land were being cleared, the trees were taking over former pasture lands.

On the farm that had belonged first to Peter Chase, then to his son Daniel, and then to Daniel's son-in-law Arad Wood, there was not much evidence of the population decline that was afflicting the community. Neighbors were fewer than they had once been, and the road had fewer occupied houses, but most of the Wood children stayed in town; just one daughter left for nearby western Massachusetts when she married. Youngest son George settled down on a farm nearby. And in 1895, a fourth owner took over the home farm. The family who bought the farm when the Woods retired were newcomers to town: German immigrants who had first migrated to Pennsylvania and then moved to Jamaica. Things seemed to prosper. They kept up the dairy operation that Arad and Fannie Wood had started in their time there. In keeping with small-town tradition, and highlighting its stability and continuity, the farm would still be known as "the Arad T. Wood place."[97]

CHAPTER 2

How Jamaica Became a Problem

The Experts in the Hills

In some ways, Jamaica was thriving at the end of the nineteenth century, its village center bustling with mills and workshops. But by the time Arad and Fannie Chase Wood retired from farming in 1890, it was clear something was wrong. The town had been losing people for decades—a fifth of its population in the 1860s alone. And it was not just the Civil War losses, although they loomed large in the public imagination. There was a brief period of stability following the war, but in the two decades between 1880 and 1900 over 40 percent of Jamaica's residents left town. By the end of the century, the town was half its former size. The low point would not be reached until 1920, when the population finally stabilized at around 560 people—a little over one third of its 1850 population.

Jamaica's decline was dramatic, but it was hardly unique.[1] Many parts of the New England countryside were experiencing similar losses in the post–Civil War years. As a region, New England's population grew by double digits in every census throughout the nineteenth and into the twentieth century, but most of that growth was in southern New England, and especially in the industrial cities of Massachusetts and Rhode Island. The frontiers of Maine continued to grow (the vast Aroostook region did not open as a potato growing region until the 1890s), but most rural areas of New England experienced more or less continuous outmigration.

Nor were such losses confined to New England. Long-settled areas of rural New York, Pennsylvania, and Ohio shared similar experiences. And these regional trends were overshadowed by the longer arc of nationwide change, as the American farm population inexorably decreased from 42.5 percent in

1890 to 37.7 percent in 1900, and then to 30.7 percent in 1910. In retrospect the process seems inevitable, part of a long national and ultimately global economic transformation from a rural and agricultural to an urban industrial society. At the time, the picture was not so clear.

Agriculture and rural life carried great symbolic weight in turn-of-the-century America. Despite their gradually declining numbers, farmers were still a large and powerful part of the body politic and were still widely believed to be essential to the prosperity and political health of the nation. Land ownership was associated not only with economic security and personal autonomy, but with the rights to citizenship and paternal authority granted to economically independent white men.[2] The political rhetoric associated with that agrarian belief may have sounded increasingly hollow in the face of the dramatic disruptions caused by the nation's rapid industrial growth, but it was still a vital part of the cultural and political landscape.[3]

New England's regional mythology added its own cultural weight. Pre-Civil War regionalists had celebrated New England as the source of the nation's greatest achievements in education, technology, and moral leadership. But in the years after the Civil War, regionalists turned increasingly toward nostalgic evocations of the pre-industrial past—"pre railroad times," as Harriet Beecher Stowe once described it. The authors of the new genre of "local color" writing turned away from the rapidly growing urban centers and their crowds of newcomers (migrants from backcountry New England as well as from Europe). They focused their attention instead on remote northern villages, infusing those places with an atmosphere of commemoration and loss. Characteristically sentimental and affectionately humorous in its early years, by the turn of the century the tone of local color fiction was becoming darker.[4] Fears for the future of rural New England were increasingly reflected in bleak literary depictions of cold weather and harsh repression. (Edith Wharton's chilling 1911 novel *Ethan Frome* is a fine example.) And those fictional accounts were increasingly joined by equally dark journalistic exposés of village decadence. In the national magazines, explanations proliferated for what was variously described as the "decline of New England," the "abandoned farm" crisis, or the "desertion of the hill towns."[5]

Our Hill Farms

Vermont's leaders certainly worried about the cultural implications of farm loss in their state: like other New Englanders, they identified rural life with the character and heritage of their people. But they worried just as much about the threat to the state's economic future. Among the New England states, Vermont was conspicuous for its relative lack of manufacturing and urbanization, and for the central place agriculture still occupied in its economic life. And it was not just a few towns in Vermont that were at risk: while the rest of the country was experiencing a population explosion, Vermont's population remained nearly unchanged for the better part of a century. By the time Jamaica hit its lowest point in 1920, a generation of politicians and agricultural leaders had struggled to diagnose and contain the state's demographic troubles, mostly without results.

Rural population loss was a challenging problem for Vermont's policymakers. Most agricultural leaders did not view it simply as the inevitable corollary of industrial progress, as later observers have often interpreted it. They feared that something was going wrong with rural communities—or perhaps the whole state. What was driving or tempting Vermonters to leave the farms? Of course, it had been clear since midcentury and earlier that many Vermonters would be attracted to new opportunities in the West, and that ambitious young people would seek their fortunes in growing cities. Direct competition from western farms was also an increasingly obvious problem, especially after refrigerated railroad cars were introduced in the 1880s. But were these the determining factors? Or was there something deeper, something deficient in the very culture or character of rural Vermonters? (Historians have struggled to answer these same questions.[6]) After all, a healthy rural society should be able to sustain itself despite the natural desire of its young people to see the world. And even if the causes of the outmigration were completely unavoidable, the consequences still had to be confronted. What did it portend for the future of the towns and farms left behind?

Faith in the Soil

The Vermont Board of Agriculture was established in 1872 in large part to diagnose and treat the problems of rural population loss and so-called farm abandonment. ("Abandonment" was a widely used term but has connotations that were rarely accurate. Most vacant farmhouses and reforesting pastures were not abandoned, in any practical sense. Taxes were still being paid; the land was usually in some kind of use.) At that point, those problems were beginning to be closely associated with the types of places Vermonters called "hill farms." That term, of course, referred to farms located at high elevations or on steep terrain, but it was not simply a geographical designation. It also referred to old-fashioned farming practices, and to farms that sent little of what they produced to market, often because they were distant from railroad connections or town centers. Most of all, the term "hill farm" had emotional resonance. It was closely associated with regional values and identity. As one member of the Board of Agriculture explained it in an 1883 talk, "Vermont is preeminently a 'hill-country.'" Of course, he acknowledged, everyone would like to have rich alluvial land to work with, but for "the large majority of Vermont farmers," the question was "how to successfully deal with land, rough and incomplete and broken." In fact, it was essential to the well-being of the entire state to solve the difficulties the hill farms faced: "If . . . our hill farms can be made to pay," then the farms "more favorably located" could take care of themselves, and "the future of successful agriculture in Vermont is assured."[7]

Over the next few decades, the Board of Agriculture considered many possible diagnoses and treatments for the rural problems that appeared to be leading to population loss. But one thing their writers and speakers did *not* do was blame the soil, the elevation, or the climate. From the early days of settlement, writers describing Vermont had been convinced that growing conditions, even quite far up in the Green Mountains, were generally good and that the elevation and climate, rightly understood, were advantages rather than hardships. These beliefs continued to be widely accepted into the late nineteenth century. Speakers and writers for the Board did entertain the possibility that the state's soils might be temporarily damaged by bad farming practices, but they adamantly defended the underlying fertility of Vermont soils—specifically including the "rough and incomplete and broken" land of the hill farms.

The board's first annual report in 1872 included the text of a lecture given by the editor of *The Vermont Farmer*, affirming that "the natural strength of Vermont soils . . . is not surpassed anywhere."[8] A few years later, Middlebury College chemistry professor Henry Seely phrased this judgment in similar terms. While Vermont soils "may for the time have reached the point at which they make no adequate return for the working," they "are never actually exhausted."[9] Seely linked that inexhaustible fertility to another defining feature of Vermont: its snow, a natural source of nitrogen gradually dissolving into the soil. "No Nile annually enriches our valleys," he wrote, "but our own ranges, year by year, send down . . . their generous contributions to replace the mineral material withdrawn by the growing crops."[10]

In fact, Vermont's cold climate in general was good for farming. Cool summer temperatures fostered rich grasslands; summer droughts seldom scorched the highland pastures. Heavy snow cover protected hillside soils from erosion. Even icy winters were a good thing, allowing for efficient transportation of goods by sled. (A snippet in *The Vermont Farmer* in the winter of 1876 read: "Do you know what this kind of winter weather is for? Why, to allow the farmers one of the best opportunities they ever had to 'skid up' their logs, and 'sled length' wood ready for the first good sleighing."[11])

So wholehearted was this belief in the soil and climate that one speaker proclaimed, "I do not believe there is a single acre of tillable land in the state of Vermont that could not be made to pay a profit." The qualifier "tillable" offered a bit of maneuvering room, but that speech ended with an even more sweeping assertion: "very few of the 35,000 farms in the state ... do not contain some land that will be profitable to cultivate if rightly managed." Of course, there was an element of boosterism in these claims; part of the job of the Board was to promote Vermont agriculture, after all. But this was also an article of faith, a matter almost of patriotism. By its very definition, the land must be able to support its people. Vermont farms might not provide their owners with great wealth, but they would always provide a livelihood in return for hard work and care.

Thus, no underlying soil or topographical deficiencies could be causing the loss of population in rural communities. And if it were not the soil, even less could it be a problem with the climate, so closely associated with the essential nature of Vermont itself. One speaker articulated what was at stake: "if there is no merit in our soil, climate or surroundings ... it is useless to try and stop the outgoing tide of emigration. We should rather hasten the departure of all that are able to go."[12] That was not a possibility the Board was willing to contemplate.

Instead, they looked for other explanations. One by one, board speakers considered most of the different theories that were popular in agricultural journals at the time. Perhaps population loss was a moral problem: the young people were leaving the farm because they were bored or greedy or overly ambitious. One speaker blamed the recent Civil War for the emerging hyper-capitalism of the Gilded Age: "a general breaking down of all moral restraint" tempted men to "run wild with hazardous and unscrupulous haste into any scheme that promised to enrich themselves."[13] Or perhaps the men of the house were too domineering: Board speakers advised fathers not to overwork their children, not to squelch their wives' efforts to beautify the household. One Middlebury College professor, for example, recalled for the Board that his own memory of farm childhood "was principally that of being tired, tired, tired," and suggested that if Vermont farms were "made more beautiful, more of our sons and daughters would return to them."[14]

Other commentators weighed economic considerations more heavily. "Vermonters... have got a living to get," one pointed out, and "however sadly and reluctantly, they must turn their backs on the green hills if they offer them nothing to do."[15] Others blamed unscrupulous western real estate salesmen and unfair railroad rates. They urged farmers to adopt new farming techniques and to respond more quickly to market conditions. Yet even these more practically minded commentators rarely suggested that farming in the northern hills was not a worthwhile enterprise—"would not pay," in the phrase they most often used. By "pay," they did not mean commercial sales alone. They meant success in what one Board member called the kind of "plain, simple, practical, legitimate" farming that would incorporate a few cash crops into a broadly diversified production plan for home use.[16] Even in remote upland towns like Jamaica, farming in Vermont, rightly understood and practiced, was by its very nature an enterprise that would "pay."

A New View

By the early twentieth century, that fundamental faith—along with all its judgments about farming practices, land, and climate—faced significant challenges. On the national level, a new approach was taking root, emerging into public discussion with the creation of President Theodore Roosevelt's 1908 Country Life Commission. Chiefly a response to the problem of declining rural population, the Commission was created in upstate New York, where depopulated towns and abandoned farms were as much a concern as they were in Vermont.[17] The Country Life Commission was headed by Liberty Hyde Bailey, founder and dean of the New York State College of Agriculture at Cornell University. Under Bailey's leadership, the Commission garnered a great deal of attention, but it was only a harbinger of a much broader "country life movement" that would become increasingly influential over the next twenty years.

The country life movement encompassed a variety of responses to a perceived crisis—or set of crises—in the countryside. Some country life advocates viewed rural depopulation as the result of fundamentally cultural problems: the social isolation of rural families; the poor quality of rural education; and the decline of rural communities in remote "hill towns." To address such challenges, they advocated construction of better roads; consolidation of rural schools and churches to promote greater social cohesion; and a school curriculum designed to encourage an appreciation of country life in rural children. Most advocates of such programs were urban professionals who feared that the traditional values of the countryside were endangered by the growth of cities.[18]

Another group of reforming professionals, however, emerged from a different background. These men and women had often grown up on farms and were educated in new scientific disciplines at the agricultural colleges. For such reformers, the single most important problem of rural life—what was really driving folks from the farm—was that farmers worked hard, but too inefficiently, for too little financial return. The goal should not be to keep everyone on farms, these reformers argued, but to raise the standards of living on farms by teaching modern techniques to those who were qualified to be farmers. Their target was the inefficient, unproductive, and outdated farming practices that appeared to be responsible for the relative backwardness and poverty of rural life. Only modernization—of crops, machinery, labor practices, finance, and even attitude—would bring better lives to farmers and keep the right people in the countryside.[19]

To those reformers, "abandoned" farms were not a social tragedy, but an expression of the natural and necessary competition of capitalism. Of course, if these new agricultural leaders succeeded in encouraging greater efficiency of this kind, many people would have to give up farming, but they generally viewed this outcome as inevitable, even desirable. As one University of Illinois agricultural scientist put it, "Many individuals will be crowded out as agriculture exacts more knowledge and skill." Ultimately, "progress is not in the interest of the individual, and it cannot stop because of individuals."[20]

Within a short time, this group of mostly college-trained researchers would take on leading roles in several emerging fields of study. At the University of Wisconsin, Henry C. Taylor established the first independent Agricultural Economics department in 1909, incorporating the new discipline of rural sociology into that program as well. Simultaneously, at Cornell University's Agricultural College, Liberty Hyde Bailey's protégé George F. Warren designed a new kind of survey that incorporated standardized questionnaires to assist researchers in gathering information from large numbers of informants.[21] Cornell students fanned out across the countryside and into hundreds of households, making detailed inquiries into farm operations, crops, expenditures, and profits and loss. With such tools, researchers could collect the kind of aggregate data that would feed the equations and charts that would in turn lead to better scientific analysis of farm problems. Their pioneering practices ultimately launched a whole generation of researchers into other parts of the countryside, questionnaires in hand.

Alongside the development of agricultural economics, the study of forestry emerged in the same years and from the same institutions: Cornell opened one of the earliest forestry schools in 1898, and in 1907 Harvard began its forestry research program at the Harvard Forest in Petersham, Massachusetts. The new field of soil science also had its beginnings around the same time. At the turn of the century, a Bureau of Soils was created at the United States Department of Agriculture. Its chief, Milton Whitney, launched an ambitious and innovative National Cooperative Soil Survey. Working closely with state agencies, federal surveyors trained in geology and agricultural chemistry (almost all from the state agricultural colleges) experimented with a variety of methods for mapping soils, with the intention of providing farmers and developers with the information they needed to make rational choices about land use. Pioneers in these emerging fields turned their attention to the related issues of land use, farm productivity, and, in the northeast, rural population decline.

Re-Making the Board of Agriculture

In Vermont, the new approach to agriculture emerged first in a series of institutional changes, beginning with modifications to the structure of the Board of Agriculture itself. In 1899, the size of the Board was reduced from eight appointees down to three. Then, in 1908, the Board was abolished altogether and replaced with a salaried commissioner of agriculture (along with a commissioner of forestry with a higher salary).[22] These changes affected not only the Board's structure, but its leadership, and ultimately its mission.

In its original form, the Board had been dominated by men with a common background: mostly well-to-do farmers, along with a handful of professors, journalists, and lawyers who had grown up on farms. (Several women did participate in the Board's programs, but they cannot usually be categorized by occupation or public position.) Appointed to the Board by the governor, most had also served in the state legislature or in other state and local political roles. One example of that old style of agricultural leadership was Board member Victor Spear, who occupied many positions of public trust over his career, from the state Senate to the Cattle Commission. Like many of his peers on the Board, Spear had a foot in two camps. Born and raised on a farm in the declining hill town of Braintree, Spear spent his professional life right next door, in a town twice the size of Braintree and decidedly *not* declining. Randolph's rapidly growing commercial and industrial village had sprung up around the Vermont

Central Railroad's White River route—a more suitable location for the merchant and political man Spear had become. In 1930, his obituary reported that the shops in Randolph closed for an hour at the time of his funeral.

Old-school Board members were clearly members of an agricultural elite.[23] But as Spear's obituary judged matters, neither his Dartmouth College education nor his business career had made him less of "a real farmer, born and bred in Vermont."[24] Although they were themselves rarely hill farmers, Board members like Spear were almost certain to cross paths with men from the hill towns in their personal, political, and professional lives. Even if no family ties drew them back, professional and political ties probably would. And after all, the smallest towns in Vermont were still represented in the state house on an equal footing with larger towns and cities. (Until the 1962 Supreme Court decision of *Baker v. Carr*, the Vermont House of Representatives was made up of one representative per town, while the Senate included both at-large members and members elected by county.) It was those old networks that would be weakened when the Board was replaced by a one-man office.[25]

A New Agricultural Establishment

Meanwhile, while the Board was shrinking, the state's bureaucratic and regulatory structure was expanding year by year, adding more salaried and credentialed staff who would enforce laws and make agricultural policy. The Board's last published report from 1908 included thirty-one pages of recently enacted laws related to agriculture. Its conference schedule that same year already included a whole array of state-employed experts as speakers, including the new state forestry commissioner, the state botanist, and two members of the state Agricultural Experiment Station.[26]

This process continued after 1908 under the leadership of the new commissioner of agriculture, whose responsibilities now included the hiring and supervising of an increasingly large paid staff: investigators who scouted out harmful insect infestations; inspectors of apiaries, nurseries, and creameries.[27] At the same time, some of the Board's former amateur affiliates were taking on greater official responsibility. The Vermont Dairymen's Association had been in operation since 1869. It was joined in 1893 by the Vermont Maple Sugar Makers Association, and in 1896 by the Vermont State Horticultural Society, all "under the patronage of the state board of agriculture."[28] In the early twentieth century, these marketing organizations began to receive independent funding from the state legislature to encourage the professionalization and standardization of Vermont's production of dairy, maple, and fruit crops.²⁹

The Agricultural College had already extended its range and influence back in 1886 with the creation of the Experiment Station for agricultural research. In 1913 the state also began to fund (and then co-fund with federal dollars) the University's Agricultural Extension Service to coordinate practical educational programs for farm households. In 1916, the Experiment Station launched its first study of the economics of dairying, and by the 1920s, it was conducting yearly studies of dairy farms in different parts of the state. John Allen Hitchcock, a graduate student in the Agricultural College at the time, conducted the first such study in 1925, employing the surveying methods pioneered at Cornell. As he explained, "The data were obtained by the 'survey' method [his quotation marks]." Investigators moved from farm to farm, studying each family's "grain bills, checkbook stubs, tax and insurance receipts" and combining them with the records kept at the local creameries.³⁰

All these new state and university functions replaced the Board of Agriculture in its former role as conveyor of expert advice to farmers. The experts themselves were changing, too.

New Leadership

For a while, local politics still held sway at the top. In 1908, when the legislature abolished the Board of Agriculture, the governor appointed to the new office of agricultural commissioner a man whose public life was nearly indistinguishable from that of earlier Board members. Orlando Martin's connections were entirely local: he came from the town of Plainfield, where he had grown up on a farm and still operated a livestock breeding business. He served in both houses of the legislature, as master of the state Grange, and for forty-two years as Plainfield's town meeting moderator. Rhetorically, Martin leaned toward the tried and true, calling on his colleagues to "put forth every effort to preserve [our] heritage from our Puritan fathers whose brain and brawn were developed among the forest clad hills of old New England."³¹

The state forester appointed that same year, however, was a different kind of specialist. Austin F. Hawes was from "away." Born in Massachusetts, he had received his bachelor's degree from Tufts University and earned his master's degree in forestry at Yale University before coming to Vermont to serve as state forester. While he held that post, Hawes was also a professor of forestry at the

Vermont State Agricultural College. By 1920, he had moved on to the United States Department of Agriculture in Washington, DC, and later became the state forester of Connecticut. Such new agricultural influencers often had few local ties and more often shared networks of expertise forged at state colleges or in other state or national organizations. These new experts were one source of new ways of thinking about Vermont's farms, soils, and rural population.

But it was not simply a case of out-of-staters bringing in new ideas: Vermonters, too, aligned themselves with the new ways of thinking. The third agricultural commissioner was appointed to his position in 1924 and held it until his death in 1947. Like the Board of Agriculture members before him, Edward H. Jones came from a prosperous farming background and had local roots: he inherited his family's farm in Waitsfield and served as the moderator of the town meeting there for twenty-five years. During his long career as commissioner, Jones would take a new approach to farming and rural populations.

The immediate economic context was an important driver of the change. The spiraling food prices of World War I gave way to a sharp drop that triggered severe agricultural depression in the early 1920s, lending credence to the idea that the problems of the countryside were rooted in a "maladjusted" and outdated agricultural sector. In a 1924 agricultural economics textbook, the prominent agricultural economist L. C. Gray explained the new diagnosis: most American farmers still harbored a lingering attachment to "self-sufficing" practices, but the national economy now required a fully modernized agricultural sector. To serve the "welfare of the nation," farmers must learn to think like business owners: keep accounts, maximize efficiency, and pay much more attention to the bottom line. In short, farms must "reach the capitalistic stage" of modern businesses.[32]

Jones took on the job of Vermont's Agriculture Commissioner in the middle of that post–World War I downturn. In his first annual report, Jones announced that despite the continuing hard times, "the most progressive farmers on the better type of farms" were actually doing fine. It was only the "inefficient farmers" (those who had presumably not yet reached L. C. Gray's "capitalistic stage") who were still struggling. That one word "inefficient" was enough to signal his allegiance to the new way of thinking, but Jones spelled it out. Reporting that those inefficient farmers were now "finding it increasingly difficult to secure a satisfactory living," and were in fact "in some measure leaving their farms," Jones suggested that "from an economical point of view," a loss of such farmers "may not be amiss." In fact, their departure might improve matters for the others who continued to farm, and perhaps help to alleviate the national farm crisis by "bringing about a better relationship between production and consumption."³³

Jones lived in Waitsfield, a town that, like Jamaica, was experiencing rapid population decline. He was no doubt personally acquainted with some of the kinds of farmers he believed should be giving up farming. He might have chosen to refer to them as "hill farmers" in his 1924 report. Although that term would still have evoked vaguely positive associations with old-fashioned simplicity and virtue, it would also have carried a whiff of judgment, suggesting outmoded, even backward ways of life. But Jones chose the term "inefficient," unleashing an unambiguously negative judgment on those places and ways of life, erasing any lingering associations with the good old days or the founding fathers. In time, that would prove to be a useful rhetorical shift, permitting Vermont's modernizing policymakers increasingly to isolate and control the problem of rural decline, assigning it to a specific set of farms and not to the whole state.

Soil Surveys

During the same years that Vermont's agricultural institutions were in flux, researchers began to appear in the state with new kinds of projects. Beginning in 1899, for example, the US Department of Agriculture's newly created Bureau of Soils launched a series of soil surveys. Among the earliest surveys coordinated by the Bureau of Soils was a 1904 study of the city of Vergennes in the Champlain Valley. A second, larger study of Windsor County soils was completed in 1916. These two surveys initiated their own approach to what was more and more coming to be seen as the "hill farm problem."

At first rather ad hoc affairs, the soil surveyors gradually built up a system of formal naming and classification that would eventually become standardized. Those technical analyses of soil types significantly altered discussions about soil fertility, in part simply by proposing new kinds of categories for it. While a Board member in 1878 could speak confidently about Vermont's inexhaustible soil—in the singular—the new studies mapped increasingly differentiated soil categories onto increasingly precise locations. The Windsor County study, for example, divided the county's soils into twelve basic categories, ranging from the best "Podunk fine sandy loam" to the worst (using a rather less technical term), "rough stony land."³⁴ Of course, farmers had already understood perfectly well that their farms included different types of soil, but the new studies employed scientific nomenclature and a hierarchical rating system to differentiate soils in absolute terms.

At the same time, responding to considerable pressure from Congress to produce practical information useful to farmers and land developers, the Bureau of Soils encouraged soil scientists to add social, historical, and economic commentary to their technical soil reports. The reports included descriptions of local farming practices, current land values, and potential new markets, and they also frequently commented on the characteristics and circumstances of rural people in the area. Those sections were never without policy implications. In long-settled regions like Vermont, the decline of farm populations was a major focus.[35]

In other parts of the country, the Bureau of Soils confronted hostile citizens (often real estate speculators or state promotional agencies) who violently resisted the description of their soils with words like "alkali" and "sandy"—deal breakers for real estate investors from Florida to Utah. (In one case, the surveyor exposed a real estate scam in the Everglades, reporting that land being sold for farms was under water.[36]) In Vermont, things did not get so heated, but the soil studies there, too, sometimes cast new doubt on the value of land. In the Windsor County report, for example, the most common soil type was said to be "Hollis stony fine sandy loam," which made up about 44 percent of the county's soil. The soil scientists described the observable features of this soil: it appeared mostly on "steeply rolling to hilly and mountainous" terrain; it was "naturally well drained," and was typically yellowish brown in color, six to eight inches deep, with bedrock at two to three feet.[37]

Probably nothing in these characterizations of the soil would have been likely to foil a real estate deal. But to these measurable features, the surveyors added several comments about the soil's historical, current, and future uses: "On the hills and mountain sides there are many abandoned farms on which the outlines of the cleared fields are hidden in the forest." And to that observation they added a different kind of assessment: "So far as can be judged, these long-abandoned farms are not capable of being successfully farmed under present conditions." What the surveyors meant by "present conditions" had nothing to do with soil. It was a reference to the use of "modern implements and machinery," which they judged would be unusable on the "steep slopes and irregular rock outcrops" associated with this soil type.[38]

That was a significant shift in perspective. Back in 1884, Board of Agriculture members had taken it for granted that machinery could not fully take the place of hand labor on most Vermont farms, but they had assumed that it was the job of the Board to help farmers find ways to work within those limits. Now the soil scientists assumed that the farmland must accommodate itself to the machinery, rather than the other way around. Implicit in their report is the judgment that a large portion of the land (over 44 percent of Windsor County, an area of the state that had previously been viewed as a highly successful farming region) was—or was now becoming—unsuitable for any kind of agriculture.

A critical moment in this process would arrive with a statewide reconnaissance survey, conducted jointly by the federal Bureau of Soils, the Vermont Agricultural Experiment Station, and the new Vermont Commission on Country Life, in the summers of 1929 and 1930 (see Figure 3). ("Reconnaissance" referred to the broad scale and generalized results of the survey.) The survey was conducted by a team of five scientists, graduates of four state agricultural colleges, who also conducted numerous soil surveys in other parts of the country.[39] At first glance, this report seems yet again to confirm the earlier consensus that Vermont had good grazing land: "The soils of Vermont are best suited to grass. Although they are not inherently as fertile as the Prairie Region or the Great Plains of the West, the humid climate prevents serious hindrances . . . from droughts and fosters a luxuriant grass cover throughout the season."[40] In fact, the soil scientists echoed much of the assessment Zadock Thompson had made a century earlier: the cold and snow were "favorable to the accumulation of organic matter on the surface of the soil . . . there is little leaching, and the summer sun is not hot enough and the warm season not long enough to destroy the organic matter."[41] Under these conditions, "the soil and climate are favorable to a rather heavy forest growth and to grass."[42]

But this moderately positive description was subverted by the implications of the details. The statewide survey used a classification system even more elaborate than those of the previous surveys, dividing all the state's soils into sixteen general categories (defined by color, texture, depth, topography, and drainage), and then gathering the sixteen soils into seven agricultural classes. At the top of the pyramid were the Class 1 soils: "smooth, well drained, highly productive land suitable for all crops." Ten percent of the total area of Vermont fell into this class. At the other end of the spectrum was Class 7, "steep stony land having little or no agricultural value." These were uplands, some in locations that were once considered good or even excellent for grazing. Class 7 soils constituted 15 percent of the state. Counted together with Classes 5 and 6, which were both designated "low productivity," the final report of the survey categorized fully half of the state's land as more or less unsuitable for farming. The town of Jamaica's soils fell mostly into the soil scientists' category 5 or 6, with swathes of "rough stony land" on the hilltops. Once described in modestly positive terms, Jamaica's soils were now categorized as mostly useless for farming.[43]

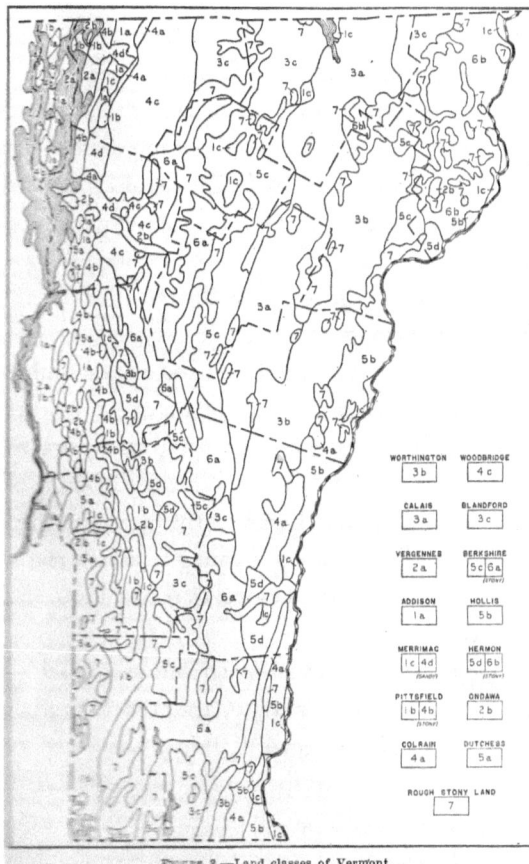

FIGURE 3. "Land Classes of Vermont." This map makes use of the detailed soil data collected by the Reconnaissance Soil Survey to establish a broad classification of soils based on their suitability for crops. Jamaica is located in the large "6a" swath of land in the bottom right—rated mostly unsuitable for agriculture. Source: W. J. Latimer et al., Soil Survey (Reconnaissance) of Vermont (Washington, DC: U.S. Government Printing Office, 1930).

Becoming Marginal

The categories devised by the soil scientists were an important part of the re-thinking of Vermont farming, but their complex nomenclature was too specialized for use as a slogan in public policy debates. Agricultural economists coined a more approachable set of terms that seemed to be more fit for general use. As one expert explained the basics:

> Land which is so poor that it will return to those who use it barely enough to pay the costs of using it is called marginal land. Land which is so poor

that it will not pay the cost of using it is called submarginal land, and land which is good enough to yield more than the cost of using it is called supermarginal land.[44]

Perhaps predictably, the term "supermarginal" never got much public use, but "submarginal" would become an important rhetorical device for agricultural modernizers. It was used all over the country to describe land that was too dry, too damp, too stony, too sandy, too salty, too forested, or too de-forested to reward the effort of the farmer. It was even used to designate soils that were simply located too far from markets.[45] In Vermont, the term "submarginal" was used above all to refer to the hill farms.

As the concept "submarginal" came into general use, the terms of debate shifted. It was not simply that the soil no longer seemed to be as good as it was once believed to be. Some agricultural experts were beginning to make a similar case about the people on the farms. Nineteenth-century Board of Agriculture writers had addressed the problem of rural depopulation by attempting to keep people on farms. They had pleaded with farmers to make the kinds of improvements that would keep their children at home on the farm; they had proposed educational improvements that would give young farmers greater pride in their work; they had pushed for improved crops, transportation, and marketing to help farmers get ahead.

Now, some experts—including the commissioner of agriculture himself—began to argue that people should not be helped to stay on the farm, but rather be encouraged to get off. If the farm was "submarginal," if the soil was Category 6, if the family was "inefficient," then, as the Board speaker had once concluded, "we should rather hasten the departure of all that are able to go." While once the state's leaders had thought the problem lay with malcontent young Vermonters who did not understand why they ought to stay at home, some had now come to believe that the problem was with rural people who could not grasp that it was time to go.

The Vermont Commission on Country Life

All the new kinds of research and analysis came to fruition in the work of the Vermont Commission on Country Life, a multiyear project that drew on the talents of several hundred of the most influential members of Vermont's professional classes. Many such state commissions and conferences were organized in the years after the national Country Life Commission set the example in

1908, but the Vermont Commission was the largest and best funded of these efforts up to that point.[46] Because Vermont's leading citizens in the 1920s still viewed the state as essentially and permanently rural, they regarded the ongoing demographic and economic losses of the countryside as an existential threat to the prosperity of the entire state, requiring an intensive inquiry into every aspect of rural economic and social life. Arranged with the cooperation of nearly every public institution in the state, the breadth of the Vermont Commission on Country Life's scope was unprecedented.

Like the Country Life movement generally, the Vermont Commission's participants were ideologically and temperamentally diverse. Some members were social workers, teachers, doctors, and ministers hoping to improve rural welfare systems, rural schools, rural health care, and rural religion. Others were professionals trained in the new agricultural disciplines: foresters, geographers, agricultural economists. Some were successful farmers; but perhaps predictably, there do not appear to have been any representatives of the hill farmers among the "200 Vermonters" who made up the Commission.[47] The important man invited from outside the state to be the director of the Commission was none other than Henry C. Taylor himself, preeminent agricultural economist and one of the founders of the field. It was Taylor who had established the first university department dedicated to agricultural economics at the University of Wisconsin; he would go on to assist in establishing the Bureau of Agricultural Economics at the US Department of Agriculture. His presence in Vermont indicated a dominant role for the new sciences in the Commission.

A Day-Long Argument

In 1929, 102 members of the Commission came together for the first time to hear the committee's report on their work in progress.[48] It was a long day. They started at 10 a.m. with a brief greeting from the governor, followed directly by the first committee report. The assembled committee members did not ask many challenging questions of one another on that day. In fact, just one matter generated any significant debate at all. In mid-morning, the Committee on Farm Production and Marketing started off that discussion.

By far the largest of all the committees, the Committee on Farm Production and Marketing already had over twenty members and would have well over fifty by the time the report was published. Its mission was to modernize Vermont farmers' production and marketing methods for the most remunerative current and potential future crops. Thomas Bradlee, the director of the state Extension Service, was the executive secretary of the committee. He spoke first, describing at length the committee's plans and its complicated subcommittee structure. Then the chair of the committee, US Congressman (and former state Agricultural Commissioner) Elbert S. Brigham, took the floor to talk more generally about the state's long-standing "farm problem." Brigham took the position that the problem was fundamentally economic. The days were gone when farmers would be contented with just a roof over their heads: "If we are going to keep our young people in Vermont, as we aim to do, we must give them a satisfactory economic opportunity. . . . not only a good home but a satisfactory money income."

On the face of it, Brigham's remark was not particularly startling. Something of a figurehead of the committee, he was evidently rehashing an argument they had all heard before—probably at Board of Agriculture meetings in the past. (The author of a *Burlington Free Press* article seemed to find the terms of debate familiar, describing the speakers as "exponents of two very different schools of thought."[49]) A clergyman responded immediately with a standard rejoinder: it was certainly *not* low incomes that explained why people were leaving the farm, he said, but isolation—not money but "the social element." A Pittsford dairy farmer named James Candon agreed with the clergyman that the problem was not a matter of money. What was needed was a better appreciation of the intangible benefits of life on a northern farm: "things that cannot be bought with dollars and cents." Of course, these were also well-worn arguments, and Congressman Brigham was ready with a retort: "we have been talking that for half a century and still our young people have been leaving Vermont."[50] Governor Weeks cut off the discussion shortly after (they were already late for lunch), but the conversation would return to the same subject in the afternoon, when the Land Utilization committee presented its report.

Land Utilization

After lunch, however, the old argument took a new turn, guided by the chair of the Land Utilization committee. The name of the Committee on Farm Production needed no explanation, but this one did. "Land utilization" was a concept used in agricultural economics. As the chair explained, it had to do with "the major uses of land areas to the end that they may yield the largest economic returns and at the same time provide a satisfactory livelihood for their occupants." Modern societies required large-scale planning and management of land use: a population "needs to be in adjustment with the economic resources of the land." In some situations, the chair explained somewhat disingenuously, it might require "getting more people on the land and in others getting some of the people off the land."

The chair of this committee was Harry P. Young, at that time a farm management specialist for the University of Vermont's Experiment Station, working on a PhD at Cornell.[51] Young presented the committee's report in a series of increasingly abrupt questions: "What can be done with farms which can not yield a satisfactory living to their occupants?" That was a paraphrase of the definition of "submarginal" farms, those which can never return enough income to cover the farmer's labor and capital input. His next question must have startled many of his listeners: "Can we recommend state acquisition of all the lands now in farms in some of these hill towns?" Perhaps realizing that the suggestion might stir uneasiness, he pressed his audience: "Are we ready to go thus far? What are we ready to do?"

Young made a cursory effort to reassure his listeners. Many of the hill farms were "clearly uneconomic," he explained, "the people [on the farms] recognize that" and were themselves "trying to get off." But his tone was strikingly uncompromising. His committee would "work up a forward looking and progressive policy" no matter what kind of opposition they encountered. After all, he concluded dismissively, "there is nothing to sob about in the whole situation." His proposal of large-scale removal of people from farms might "tread on some of the ideas of the native Vermonters" (he was from Pennsylvania), but that "might be good for it [sic]."[52]

Afterward, Young attempted to soften his tone a bit. The secretary of the commission sent him her minutes of the meeting for his approval before she published them, and he took the opportunity to remove that harsh-surrounding statement: "There is nothing to sob about."[53] But it was too late for the afternoon's discussion. Things were becoming a little personal. Offering a cautionary tale about the plight of a farmer in the lower Champlain Valley (where the clay soil made roads impassable for part of the year), Young argued that those bad roads were responsible for farmers' sons rejecting a future on the farm. In an attempted joke referring back to the morning's discussion, Young quipped that the real reason the farmer's boys had gone off to town was because the farmer had neglected to speak often enough about Vermont's beauty, "as John [sic] Candon would have told them."

As it turned out, Young did not have the last word. Jay J. Fritz, member of the Forestry Committee and the forester of Middlebury College, seemed more than a little startled by Young's proposal: "I have only known Mr. Young for a short time, but during that time I have found that he is absolutely fearless. When he gets an idea he will spring it on you in a cold blooded manner." Fritz agreed that more land should be in forests, but Young's proposal of wholesale abandonment of the hill towns was too much for him: "The suggestion of buying up all the civilization in the town, removing it and then throwing it into wild land with no town expenses, I do not believe would apply or could be applied to a great many towns in the state of Vermont." (Apparently recovering from his surprise, Fritz later walked back his resistance slightly, agreeing with Young that in some situations it might be a good idea for the state to buy up land for reforestation or development for tourism.[54]) After Fritz sat down, the chair decided to move on.

Six more committee reports followed in quick succession, generating little discussion.[55] But the last presentation that afternoon came from the Committee on Traditions and Ideals, headed by Arthur Peach, English professor at Norwich University (then known as the Military College of the State of Vermont, located in the hill town of Northfield).[56] Peach began by situating himself and his committee in direct opposition to the prevailing tenor of the day's meeting: "We have listened to-day to many practitioners of . . . the 'dismal science of economics,'" he began. Peach recognized the hypnotic force of all that aggregate data, but he was determined to challenge it directly. His committee's work, he explained, was specifically to "study the intangible values that no graph by the expert in economics can readily reveal."[57]

Looking back to the morning's discussion, Peach acknowledged Congressman Brigham's original point that "many a Vermont hill-side farm is an uneconomic proposition," but he maintained that economics should not be the only gauge of value. Alarmed by what he had heard about plans to "buy up all the civilization" in the hills, Peach pleaded with his listeners to "take into consideration . . . the inner desire in the heart of a man that makes him cling to that hilly home . . . and . . . to the mountain hamlets." In keeping with his committee's charge to defend Vermont's "Traditions and Ideals," he articulated the values he believed kept Vermonters on the hill farms: "Independence of spirit, love of the hills, some sense of freedom that far hill vistas give." In a direct appeal to his fellow commission members, he asked, "these have definite values in a scheme of life, have they not? Though they bear no cash dividends in the marketplaces of men." If those values were lost, it would indeed be something to sob about, despite Young's flippant remark earlier in the day.

Peach concluded, moreover, with a very different kind of economic argument. The hill farms provided something more important than a beautiful view or a sense of freedom: it was the basic security of "a home from which a man can neither be starved nor frozen as long as a crop will grow and wood burn."[58] This part of Peach's argument would take on more force in a year or two as the Great Depression settled over the nation. Now, it seemed to go for little.

The Thirteen-Town Study

The real work of the VCCL, despite all the talk, was not done by the committee members but by the people hired to conduct surveys. The Summer Residents committee commissioned a study of households that took in summer boarders. The Committee on Religious Forces conducted an elaborate survey of the churchgoing habits of rural Vermonters. The Committee on Home and Rural Life studied standards of living and household expenditures. The Committee on the Human Factor commissioned a study of "Agricultural Populations in Select Vermont Towns" to explore the impact of immigrants on rural Yankee "stock."

The Land Utilization committee did not have to pay for its own survey: they made use of a comprehensive multidisciplinary study sponsored cooperatively by the federal Division of Land Economics, the Vermont Agricultural Experiment Station, and the Vermont Forestry Service. The study was conducted in the summer of 1929 by a team of two foresters, two agricultural economists, and a geographer. One of the team members, agricultural economist Lemuel J. Peet, prepared the report, which he submitted as his thesis for a master's degree at the State Agricultural College. When the report was published in the Vermont Agricultural Experiment Station bulletin in 1933, it listed Peet as a coauthor with Claude F. Clayton, then Senior Agricultural Economist at the federal Division of Land Economics. By that time Peet had been hired as an Assistant Agricultural Economist in that office.

Peet and his colleagues conducted an intensive study of thirteen hill towns scattered along the ridge line of the Green Mountains. The study, as Peet explained, originated in response to "the uncertainty in the minds of both residents and non-residents of these towns with regard to the proper use of the land." The purpose of the study was to discover and to explain the "proper utilization of the land in these towns" and the many other towns like them: a total of eighty-seven towns the teams identified as similarly troubled.[59]

The five investigators studied nearly every aspect of the land and of the economic lives of the residents of the thirteen towns (see Figure 4). The foresters mapped the advancing forest cover; investigated the quality of the existing woodlands; calculated the amount of maple syrup produced and the number of cords of wood. The geographer mapped the farms, the acreage in different soil classes, the production of the farms, and their proximity to roads, schools, doctors, and town centers. The team chose 162 representative farms to study in depth. For those households, the researchers gathered information about yearly budgets, family income, expenses, and savings. The farm economists visited the town clerk's offices and copied the property tax information on the grand lists; they recorded family sizes, educational attainments, ages at marriage, and outmigration patterns. Finally, they determined a "desired farm income" for each household and attempted to calculate what it would take to help those farms attain the income for a proper "standard of living."[60]

The resulting report was an exhaustive evaluation of social, financial, and land use problems in the hill towns. Those problems were clearly acute: the towns were experiencing rapid population loss; rapidly declining numbers of farms and open farmland; low incomes and few non-farm jobs or educational opportunities. Yet the results were not all gloomy. One of the team's most striking discoveries was that, while the land was indeed stony and the climate cold, those factors were not the important deterrents to farming some policymakers had anticipated. The short summer seasons limited the production of corn, Peet wrote, but "hay, so essential to a dairy farm, yields well on this land, even when cropped continuously for long periods." Pasture was ample to support cattle in summer and the farmers of these hill towns understood cattle and dairy production. Nor was the problem low fertility. As reported in the Vermont Commission on Country Life's monthly newsletter in July 1930: "Contrary to general opinion it was not found that hill farms had been deserted because of loss of fertility of the soil."[61]

Yet the hill farmers did face daunting handicaps. The soil might be fertile, but it was stony, and the fields were bound by old stone walls that could not accommodate machinery suited to larger fields. It was reasonably good dairy country—lying in the "so-called hay and pasture belt"—and it was close to well-established markets for dairy. But lowland farmers were currently being forced

FIGURE 4. "Farms Studied in Detail," Map of Granville and Warren, Vermont, two of the thirteen towns studied by the team led by Lemuel J. Peet. The map is an example of the team's exhaustive and meticulous research. The locations of the farms studied in detail are laid over the contour maps generated by the United States Geological Survey, so that a close reader may compare the lay of the land on each farm with its status in 1929. It marks current use for each farm, ranging from those still operated to those that had been "abandoned" for ten years or more. Source: Lemuel J. Peet, "Problems in Land Utilization in the Hill Towns of Vermont, Based on a Study of Thirteen Towns in 1929" (master's thesis, University of Vermont, 1930), 15. University of Vermont Archives, Silver Special Collections Library, University of Vermont.

to shift from cream (for butter production) to fluid milk, as low-cost western butter was pricing New England butter out of northeastern urban markets. Hill farmers, as Peet explained, either have not "effected the change or [have] done so only recently." Perhaps it was "because of reluctance," but he thought it was "largely because of inability to meet the requirements of the fluid milk market." While cream for butter making had typically been transported by wagon to local creameries or to the railroad depot, fluid milk was now increasingly being transported by truck. Many hill farms were located on distant and poorly maintained back roads, too remote to make trucking feasible, particularly because shipping fresh milk was so time sensitive. As Peet's team judged, it was a market in which hill farmers would be "hopelessly outclassed."[62]

The thirteen-town report showed clear signs of the investigators' sympathy with the residents of the hill towns. Lemuel Peet was himself the son of a farmer in the town of Cornwall (a town that was also losing population, although it was not one of the eighty-seven towns his study documented). He noted several times that the poverty of the hill farmers was not due to their lack of intelligence or work ethic, but to a cycle of events that had worked against them. "The people living in these towns are alert, and progressive to the extent of their means." He particularly praised their willingness to sacrifice to provide education to their children.[63]

Perhaps because he sympathized with their situation, he found himself reluctant to make the kinds of sweeping recommendations his sponsors were probably expecting, concluding that "it is difficult to say what the population might best do." Perhaps the people were better off in their mountain homes than anywhere else. Indeed, their "character might be such as to make [them] unfit for any other environment." In that case, this idealistic young economist concluded, a correct understanding of the principles of land utilization would lead to the conclusion that they should stay where they were: "the people are located at present on land which will give to them and society the largest amount of the goods and satisfactions of life."[64]

Notably, Peet did not recommend wholesale removal of the hill farm residents. (That would be left to others.) Most farmers had little in the way of savings other than the value of their homes and farms, and thus would find it prohibitively difficult to leave town even if they wanted to—which was far from clear, despite H.P. Young's assertion that they wanted "to get off." Those who had stayed in the hills up to this point were generally those who had access to inherited land or businesses. Just as Hal Barron reported in his 1984 study of an earlier generation in the town of Chelsea, the thirteen-town study found that most sons of farmers—that is, most men who stood a chance of getting a farm—still lived in town or nearby in 1929.[65] And as Barron found in Chelsea, Peet suggested that what the towns needed most was non-farm jobs (in these hill towns mostly in lumber mills or in the woods cutting timber).[66]

In the coauthored 1933 Agricultural Experiment Station Bulletin, the list of recommendations was more detailed and the language more bureaucratic, but the two authors still stayed away from any sweeping recommendations of removal from the hill farms. It did not seem likely, Peet and Clayton wrote, that the present inhabitants of the hill towns could be tempted to leave. There were no businesses now in place or likely to be developing that could reasonably be expected to "induce" the hill farm people "to abandon houses, cleared land, woodland, public buildings, roads and other tangible evidences of wealth or assurances of security." Any lasting changes to the hill towns would require state intervention and significant funding. Perhaps the state could create zoning to prevent people from farming in remote areas, but zoning might be perceived as coercive, and the team predicted it would be "repugnant" to many people. They actively dismissed the idea of large-scale state acquisition of land, since there was no urgent state interest to be maintained, nor could they justify it by unpaid taxes, since hill farm owners almost always kept up with their taxes. In the end, the best path seemed to accept and encourage the reforestation of the region and promote more forestry and tourism as sources of additional sources of income for those who stayed in the hill towns.[67]

How Politicians Use Academic Research

Those were not the results the Land Utilization committee was looking for. The thirteen-town report provided the committee with a wealth of data and no shortage of carefully worked out recommendations. The committee used some of the data but ignored most of the recommendations. (In fact, the Committee on Home Life used the thirteen-town study more effectively than did the Land Utilization Committee. That was the work of Marianne Muse, a home economist who will appear in chapter 5.) Instead, they used the thirteen-town study to support policies that were in many ways at odds with the study's findings, and that would require sweeping state intervention into the hill farms. Bullet points included: "state acquisition of farmlands suited only to forestry;" "revision of the town organization to reduce the tax burden" on timber investors; and "encouragement of lumber companies to acquire large holdings of timber land in these towns." The committee's attention to the needs of large landowners was evident: less so, their attention to the needs of hill town residents for jobs. They looked forward to a day, "after the schools are closed and many roads are abandoned" in the hill towns, when the taxes on large landowners would decline and "forestry [would become] more profitable as an investment for private funds."

The committee concluded its report with a recommendation that "the responsibility for the solution of the hill town problem should be assumed by the state." The state should establish a permanent land utilization commission to "work out a solution for each town."[68] That commission should include the state forester, the commissioner of agriculture, an agricultural economist from the Agricultural

College, and "two other men chosen to represent the lumbering, wood working, and recreational interests." Notably missing from that list: hill farmers and the people of the hill towns. The Peet team estimated that the conditions in the towns they studied were typical of at least 87 of the 251 towns in the state. If the Commission intended to recommend the removal of all the people from even a few of those towns, it would be a massive project indeed (see Figure 5).

Jamaica Meets its Own Experts: The Eugenicists

The town of Jamaica was not part of the thirteen-town study, perhaps because it was already under consideration for a different investigation. The Vermont Commission on Country Life was profoundly influenced by one field of study that may seem at first glance to be unrelated to the new agricultural sciences: the state's increasingly high-profile eugenics movement. In fact, the Country Life Commission itself was in some ways a kind of auxiliary of the eugenics movement. The founder and director of the Eugenics Survey of Vermont, University of Vermont Professor Henry Perkins, played a formative role in creating the Commission, seeing it as an opportunity to give his eugenics efforts better press and a greater field of endeavor.[69] As he presented the case for creating the separate new commission to the board of the Eugenics Survey, he argued that rural problems were by their nature eugenical: "the problems of delinquency and deficiency in Vermont are very largely affected by, if not the results of, rural conditions."[70]

Within the committee structure of the Country Life Commission, the eugenicists spoke officially through the Committee on the Human Factor. And it was under the auspices of that committee, chaired by Middlebury College president John Moody, that a different investigative team was assembled in the year the thirteen-town study was under way. Instead of men trained in geography, agricultural economics, and forestry, this study would employ women trained in the survey tools of the Eugenics Survey. Based on that team's study of three towns, it would be the Committee on the Human Factor that would ultimately make the most radical recommendation of all for the hill towns. That committee would propose that the state "encourage people who live on marginal land to move to the more progressive communities of the State"—and that it should accomplish that "by taking over all marginal land."

What was it about the eugenics team's experience in Jamaica and the other two towns that led them to this sweeping conclusion? (It cannot have had much

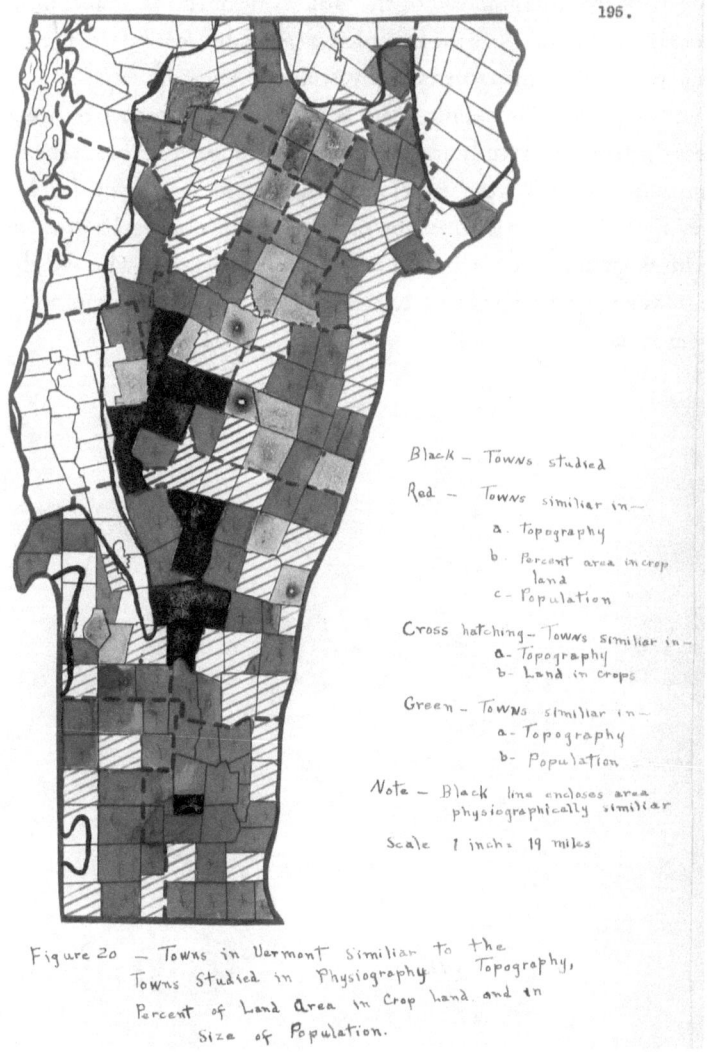

FIGURE 5. "Towns in Vermont Similar to the Towns Studied in Physiography, Topography, Percent of Land Area in Crop Land and in Size of Population." This hand-colored map indicates the eighty-seven towns the Peet team identified as having problems similar to those of the thirteen hill towns they studied. The implication was that all these towns might require state intervention. Among the 87 towns were Jamaica and all its adjoining towns. Wardsboro, the town adjoining Jamaica to its south, was one of the thirteen towns studied, and is shown here as the southernmost town colored black. Source: Lemuel J. Peet, "Problems in Land Utilization in the Hill Towns of Vermont, Based on a Study of Thirteen Towns in 1929" (master's thesis, University of Vermont, 1930), 15. University of Vermont Archives, Silver Special Collections Library, University of Vermont.

to do with the "marginality" of the land mentioned in the recommendation. There is no evidence that the eugenicists calculated, or, indeed, even knew how to calculate, hill farms' returns on investment.) As it happens, that question can be answered in some detail. There is one crucial difference between Lemuel Peet's thirteen-town study and the three-town eugenics project: the eugenicists' manuscript records were preserved in full, including detailed field notes for every household in the three towns. These records reveal a great deal about the investigation, the town, and even about the women who did the work of the survey. Chapter 3 will turn to an account of the investigators and what they saw in Jamaica.

CHAPTER 3

The Eugenics Survey

A Diagnosis and a Proposed Solution

Just after Christmas 1930, the *Brattleboro Reformer* published a notice in its society column. Along with the column's news about local citizens who had gone to Florida for the winter and children who were home from school, the paper reported that the survey being conducted in the outlying town of Jamaica was now complete, and the investigators, Miss Anderson, Miss Choate, and Miss Rome, had returned to Burlington. The three women were reported to be "very much pleased with their reception in Jamaica and deeply appreciate the courtesy which has been shown to them."[1] (The Jamaica correspondent would be less pleased with the team a few years later.)

Elin Anderson, Marjorie Choate, and Anna Rome were the field investigators assigned by the Vermont Eugenics Survey to conduct studies of population decline in three sample towns: Waitsfield, Cornwall, and Jamaica. Like many eugenics organizations, the Eugenics Survey of Vermont's staff was composed mostly of highly educated women. Elin Anderson had recently been hired as the chief investigator of the Eugenics Survey, and over time she would play an important role in that organization. Born and raised in faraway Manitoba, Anderson held a master's degree in social work from Columbia University.[2] Marjorie Choate was a graduate of Syracuse University; after her year's work in Vermont, she went on to a long career collecting data in a variety of private and public settings.[3]

Anna Rome was the Eugenics Survey's office secretary. The Eugenics Survey's director, Henry Perkins, referred to her as "one of the two girls in the office" in a letter to the Advisory Committee of the Eugenics Survey, but reported that "she has been tried out in the field as assistant investigator and has done well although her work as office secretary has been her only training."[4] In fact, Rome had graduated from Burlington High School with a degree in "commerce" at a time when only a fifth of American women completed high school. The daughter of Jewish immigrants, she had already come a long way when, at just twenty years old, she was assigned to work on the surveys.

Technically, the Migration Study, as it was called, had two institutional homes: the investigators were hired and paid by the Eugenics Survey, but their findings were employed both by the Eugenics Survey and by the Country Life Commission's Committee on the Human Factor. In practice, there was little distinction between those two organizations: the Committee on the Human Factor was nothing more than the Eugenics Survey operating from within the Commission. Either way, the Migration Study reflected the priorities of University of Vermont Professor Henry Perkins, the founder and director of the Eugenics Survey.

Perkins had originally founded the Eugenics Survey in 1925 with the goal of convincing the public that eugenical solutions—particularly sterilization, but also incarceration and other types of controls—were appropriate ways of dealing with the state's social problems.[5] In keeping with that goal, the Eugenics Survey's first major undertaking had been to create genealogical records of "low grade" families that would demonstrate the destructive genetic patterns of "defect, degeneracy, and dependence" Perkins believed were lurking in the population. The Survey's first field investigator, Harriet Abbott, was essential to that project: Abbott was "borrowed" from her job with the Vermont Children's Aid Society, and she brought with her the extensive data she had gathered about the people who received assistance from that organization. This information, along with additional data she was able to gather while working for the Eugenics Survey, became the basis for the detailed genealogical "defect" charts that Perkins used to demonstrate the need for intervention in the state's eugenical problems.

By the late 1920s, however, Perkins was attempting (in common with many other eugenics groups) to project a more popular and upbeat message. The first and second annual Eugenics Survey reports had focused largely on family defect studies (including the infamous "clock dial" charts tracing defects through the generations), but the third report in 1928 announced a turn toward studies of the so-called "better branches" of families. As Perkins explained, "it has been brought to my attention that some members of the Advisory Committee have been a little dubious about our placing so much emphasis upon low grade families." Now, he promised, the organization would turn its attention to "families of much higher grade levels, people who have made contributions to their communities."[6] The "better branches" study was something of a fig leaf: Perkins did not actually back down on his studies targeting vulnerable "low grade" populations. In that same year, the chief investigator of the Eugenics Survey was engaged in studies of two institutionalized populations: women imprisoned on "morals" charges and children on the waiting list for admission to the Brandon School for the Feeble-Minded.

But Perkins had long been envisioning an ambitious project that would integrate eugenics into the progressive mainstream: a broad-based and multi-faceted "comprehensive rural survey." It would be designed "not solely to gather facts" about rural population loss, but also to solve the problem: "to check the loss of population in rural areas of Vermont, and similarly of every state in the union."[7] That bigger project became the Vermont Commission on Country Life. Its work would take three years, involving hundreds of Vermonters organized into sixteen separate committees, and culminating in 1931 with a comprehensive report: *Rural Vermont: A Program for the Future, by Two Hundred Vermonters*.

The Eugenics Survey's influence was felt in every part of the Commission on Country Life. It is evident in *Rural Vermont*, for which the Committee on the Human Factor wrote the first substantive chapter, titled "The People," and opening with the question: "What is happening to the old Vermont stock?"[8] But the eugenical mission was also evident in the Citizenship Committee's exploration of the possibility of restricting the franchise to educated native-born voters, and the Committee on Summer Residents' questions about the "quality" of summer visitors. And it had a profound influence on the Commission's approach to the problem of rural population decline.

For Henry Perkins, the Migration Study of Jamaica, Waitsfield, and Cornwall was a critical part of the plan to place eugenical studies under the more broadly acceptable umbrella of the Country Life Commission. The study was organized around a research question that was remodeled to fit within a eugenics framework, focused not on the declining *number* of people leaving these small communities, but on "the kind of people" who were leaving.[9] Still, Perkins was aware of the need to protect the Commission from the publicity associated with the harsher aspects of eugenics.[10] Studying the eugenic "decline" of the people who stayed behind on Vermont's hill farms would require careful management.

The Decline of the Yankees

In recent times, the eugenics movement has become infamous primarily for the pervasive harm it has done to Black and Indigenous people, and to other stigmatized racial and ethnic groups. And indeed, eugenicists spearheaded all kinds of successful racist ventures: from the successful effort to cut off the flow of immigrants from racially "undesirable" parts of eastern and southern Europe in the 1920s; to the widespread state sterilization laws disproportionately used against people of color throughout much of the twentieth century. At the same time, eugenicists were often engrossed with what they viewed as the genetic decline of white populations. The earliest family studies investigated rural white families like the pseudonymous "Jukes," "Kallikaks," and "Nams," searching for explanations of how these families' undeniably "good" genetic inheritance could have been subject to such dramatic deterioration.[11]

In Vermont, it was, similarly, the decline of the state's distinctive "Yankee" genetic inheritance that seemed to pose the most profound threat to the future of the region.[12] Thus the fact that almost all the people who were interviewed in the Migration Study were white—and not simply white but descendants of the New England ancestors so admired at the time—did not protect them from eugenical investigation. Indeed, it was the very factor that provoked the eugenicists' interest. Building on a popular assumption that it was always the "best and the brightest" who left the hill farms and towns, eugenicists inferred that outmigration and genetic decline were inextricably linked. It is true that Jamaica was not quite as exclusively "Yankee" as people at the time imagined. Fifteen percent of the Jamaica interviews contain a reference to someone in the household whom the investigators believed to be of non-Yankee descent (often by way of a distant ancestor).[13] But above all, "Yankees" were the problem race for Vermont eugenicists, just as hill farmers were becoming the problem class.

Yet Perkins had already encountered resistance both private and public to the idea of applying his style of "negative eugenics" to broad-based studies of rural Yankee Vermonters. At the first meeting of the Executive Committee of the Commission on Country Life in May 1928, for example, Commissioner of Education Clarence H. Dempsey remarked that there was "one statement of Dr. Perkins I rather doubt;" it was the contention that the rural population was "falling off . . . not only in quantity but in quality." Dempsey acknowledged that such a decline might be found "here and there, but I doubt whether it is generally true."[14] The residents of Vermont's hill towns were not downtrodden peasants; they possessed both local power and social standing, and as a group they retained an important cultural significance in the eyes of other Vermonters.[15] Because of the way the state legislature was apportioned, they even held enough political power to make some people nervous: one member of the Land Utilization committee pointed out (with some exaggeration) that "hill towns, under the present system, have greatly magnified political powers and really govern the state."[16]

Whether the residents of the hill towns held that much power or not, it would not do to make wholesale diagnoses of racial degeneracy or genetic "defect" in such "old stock" rural communities. Indeed, by the time of his 1932 report to the Third International Congress of Eugenics, Perkins had publicly backtracked on the whole idea, explicitly discounting the widespread and long-standing notion that the "best and the brightest" had left the state and acknowledging that "many men and women of the highest quality remained" in Vermont.[17] As it happened, then, the studies of Jamaica, Waitsfield, and Cornwall would make comparatively little explicit use of the framework of "defect, delinquency, and dependency" that had been so fundamental to the earlier Vermont eugenics studies.

The new field investigator Elin Anderson also likely played a role in formulating a different approach. In fact, Anderson would later help to shift the work of the Eugenics Survey away from its focus on genetic determinism and toward a more nuanced consideration of environmental causes of social problems. Her greatest contribution to the Eugenics Survey's research would be a groundbreaking study of the many different ethnic groups in Burlington, Vermont, a direct rebuttal of former eugenical assumptions about ethnic and racial hierarchies. (Published in 1937 by Harvard University Press as *We Americans*, it won a prize for the best book published that year to promote inter-racial understanding.[18]) The three women who conducted the 1930 investigations may have been trained in eugenical inquiry, and the survey forms they used had clearly been designed by eugenicists, but the Migration Study they would produce would not be quite the usual eugenics fare.

The Hypothesis

From the earliest stages of planning, the Country Life Commission leadership steered the Migration Study in a clear direction—so clear, indeed, that the results would amount almost to a set of foregone conclusions. At a 1928 meeting of the

Country Life Commission's advisory committee, Martha Wadman, then the chief field investigator of the Eugenics Survey, took suggestions about possible towns for the planned study.[19] (Wadman was a field investigator of the Eugenics Survey for only a year and did not end up participating in the studies.) Three towns were to be chosen to represent "good, high average, and poor rural areas." They were to be "typical of the greater number of Vermont towns:" that is, fewer than a thousand inhabitants and "composed mostly of old Vermont stock." The study would focus specifically on Yankee farm communities. (Communities with many people of French Canadian and other immigrant backgrounds were to be the subjects of another set of studies, by geographer Genieve Lamson.[20])

The members of the advisory committee made several suggestions. Journalist Walter Hill Crockett recommended that the investigators choose the town of Waitsfield as an example of a "high type" of town, characterizing it as an "unusually intelligent community." Others at the meeting enthusiastically agreed that despite its dramatic population losses, Waitsfield was a place that was adapting well to modern times. (It so happened that one well-known native of Waitsfield was Agriculture Commissioner Edward H. Jones—the one who had been so nonchalant about the loss of "inefficient" farms a few years earlier. He was a member of two large and influential commission committees.) Next, the town of Cornwall was proposed as a "high average" community by Thomas Bradlee, director of the Extension Service at the University of Vermont and a member of the Commission's Executive Committee. Bradlee described Cornwall as a place with "good farmers and good people," and suggested it would be interesting "to know why this town has kept up" despite its population loss. Elwin Ingalls (leader of the 4-H programs at the state Extension Service) commented that there were "some French people on the farms" in Cornwall, but they were mostly second-generation and had partially assimilated into the community.[21] (Lemuel Peet, the leader of the 1929 thirteen-town Land Utilization study, had grown up in Cornwall on his father's large and thriving farm.)

It was William Frazier, a Congregational minister and member of the Committee on Religious Forces, who proposed that Jamaica be considered as the representative of towns in "poor" condition. At this point, neither Jamaica nor any of its surrounding towns was represented on the Country Life Commission. (Later, aware of the absence of voices from the hill towns, the Commission's executive committee belatedly added a few additional members, including one from Jamaica.[22]) Frazier was not from Jamaica, either, but he had at least visited the town, and was able to explain to Wadman what she should expect to find there. In his view, Jamaica's most significant problem was not its striking loss of population, which, after all, it shared with many other "better" towns. It was instead a sort of collective character flaw (if not a genetic defect): "The impression, on entering the town, is that of lack of initiative." Frazier thought that the older people showed more enterprise than the younger ones, and he assumed it was because the "best and the brightest" were leaving the countryside, a "weeding out process in which the better ones have gone." The usual metaphor for this commonly held belief was "skimming the cream." ("Weeding out" did not make much sense if the weeds were supposed to represent those who were leaving.) But Frazier's emphasis on Jamaicans' lack of "initiative" was a refinement of this old notion, and his way of looking at things would turn out to have a powerful influence on the study.

Elin Anderson replaced Wadman as the chief field investigator for the Eugenics Survey just in time to take charge of the Migration Study. Her analysis of Jamaica and the other two towns would play a critical role in formulating the Commission on Country Life's overall diagnosis and proposed treatment of rural population decline in Vermont. From the outset, Frazier's comment appears to have shaped the focus of her inquiry. But Anderson would ultimately conclude that the lack of "initiative" Frazier diagnosed in the current residents of Jamaica was not the result of all the "best and brightest" people departing (or being skimmed or weeded) for the cities and the West. Perhaps following her own ideological or personal predilections, she identified a different process at work.

Anderson came to town with a fully formed expectation of what she would find in Jamaica. But her personal reaction to the place also seemed to play a role in her analysis. As she reported it, her first encounter with Jamaica was a disheartening experience. Driving "eastward down the mountain ridge," the first thing they encountered was "a home—bleak and weather-beaten." (Evidently, they drove into town from the west, coming across the mountains from Manchester. That route would have exposed them first to the wildest and most de-populated parts of town.) Foreshadowing by a few years the Great Depression's Farm Security Administration photographs of down-and-out America, Anderson created a vivid word picture of the isolation and poverty she saw: a dooryard "strewn" with "parts of implements" and "sticks of wood," chickens "roaming," a cow grazing, a small garden, and an "old Ford car." Surrounding this, she wrote, "are miles of wilderness."[23]

Perhaps the people who lived in that house would have been surprised to hear it described in such dismal language. One might imagine a different reading of the landscape: evidence of access to decent meals (the chickens, the cow, and the garden), for example. If the inhabitants of the house felt lonely out in the wilderness, they did have a car to take them to town, after all—a modern convenience that many rural households did not have in those days—and they clearly lived on a main traveled road.

Looking back a few years later, indeed, the local newspaper correspondent for the *Brattleboro Reformer* made a well-founded accusation that the whole study had been just a case of cherry-picking the evidence. "You know how easy it is to take pictures of abandoned buildings on some road that hasn't been used for years ... Well, that's what they did." The team had already picked out a "beautiful" town and a "busy" town, she explained, and now they needed a town that was in bad shape. The result was that "Jamaica, which is one of the nicest towns with as nice people as you will find anywhere, was made to looking like the jumping off place."[24]

At least at first, Anderson clearly did find the town bleak and depressing. Yet at the same time, she could not escape the sense that Jamaica was a strangely appealing place—perhaps almost as "nice" as the local correspondent felt it was. The pseudonym the survey chose for Jamaica was "Sylvania," evoking the atmosphere of a fairy tale. (In contrast, "Beaufield" and "Pomona"—the pseudonyms for Cornwall and Waitsfield—seem designed to emphasize their productive agricultural traits.) Describing the town's lost-in-time ambience, Anderson wrote: "Sylvania is another world." She confessed she found "a charm about Sylvania that is not found in the other two towns." In the few months she had visited the place, she almost felt as if the atmosphere was causing her to lose her own initiative: "one becomes so imbued with its spirit of carefree contentment that the hustle and hurry on the other side of the mountain seems useless and vain."[25] For all the world as if she were an American tourist in a village in Italy or Mexico, she wrote: "Perhaps the charm of the place is best expressed by one of the citizens who explained—'It is always afternoon here.'"[26]

In the end, Anderson solved her interpretive problem by linking that peaceful atmosphere directly with the town's decline. The people of Jamaica lacked "initiative," not because their genes were deteriorating, but because the environment of Jamaica could not provide the necessary "stimulus"—the challenges, hardships, and competitive incentives—to bring out "the best inherent capacities of the citizens" and to encourage "the children to greater effort." Of course, Anderson was not referring to the natural environment. The long winters and rugged landscape presented the residents with plenty of challenges and hardships. What she meant was that the community in Jamaica somehow did not exert enough social pressure on residents to compete and struggle for success. In a way, she was saying life was too easy there, the residents too comfortable with their lives.

Jamaica was not regressing because of a deteriorating gene pool. It was the other way around: the town's social environment was causing the deterioration of its human "stock." In fact, Anderson argued that the high-quality genetic material of these Yankee families was susceptible to decline *only* under the social circumstances she found in places like Jamaica: "poor isolated communities where the potential capacities of the people are not challenged into use." Their genetic heritage had not been irretrievably damaged. Their inherent talents were only "dormant"—awaiting the stimulus of a place where they would have to struggle to survive. Of course, if they remained in carefree Jamaica, where it was "always afternoon," who knew what damage might occur.

The Remedy

Anderson's report appeared in 1931 as the *Fifth Annual Report of the Eugenics Survey of Vermont*: "Selective Migration from Three Rural Vermont Towns and its Significance." It presented her ingenious diagnosis, along with a proposed remedy for the problems of the hill farms that was both more unexpected and more radical than the solutions proposed by other experts who were studying the problem of the hill farms.

Anyone who had been paying attention to the work of the Commission would probably have been expecting the Land Utilization Committee to recommend the widespread removal of hill farm populations. Harry P. Young, the chair of that committee, had in fact already made such a proposal before any research at all had been completed, back at the first meeting in 1929: "Can we recommend state acquisition of all the land in farms in some of these hill towns?"[27] In the end, the committee's final report was slightly more restrained. It did not use the word "all," but it did recommend "state acquisition of farm lands suited only to forestry," along with the "encouragement of lumber companies to acquire large holdings of timber land in these towns." The Forestry Committee, similarly, recommended lowering taxes on potential timber investors. As Mortimer Proctor, the chair of the Forestry Committee, had explained in the 1929 meeting: "forest land and farm land won't mix." The state should set aside areas for timber investment and "permit whatever farming population exists therein to depart."[28]

On the face of it, Anderson's proposed remedy for the hill farms' problems was much the same as the recommendations of the other two committees. The state, she wrote, should "encourage people who live on marginal land to move to the more progressive communities of the State." But unlike Harry Young, Anderson *did* use the word "all." In full, her recommendation was that "the state encourage people who live on marginal land to move to the more progressive communities in the state *by taking over all marginal land* [my emphasis]."[29]

Perhaps a greater understanding of the limits of state power restrained the Land Utilization Committee: Anderson may have had little notion of what "taking over *all* marginal land" would entail. But that "all" did represent an important difference between Anderson's proposal and those of the other committees. The Land Utilization and Forestry committees were closely aligned with the interests of the forestry industry and to some degree with the tourist industry, as the language of their recommendations would have made clear to anyone who did not already know it. Their proposals were intended both to protect Vermont's forest resources and to bring more business to the state. Both of those objectives would require large-scale efforts, but Anderson's proposal was more far-reaching because the urgency of the problem she perceived was so much greater: in Anderson's words, a "challenge" had been "flung to the state." If Vermont wanted "its future citizens to have the same fine qualities of character that marked the early builders of the State," she wrote, it must "prevent deterioration from taking place." That would require "a social environment that will continue to bring out all the fine qualities in the character of its people." In the case of the hill towns, the social environment was beyond repair: the people living there would have to be moved out of the hills to "more progressive communities," and they would have to be "encouraged" to move by a state takeover of "marginal" lands.[30] State acquisition of scenic woodlands could be done one piece of land at a time; lowering taxes on woodland could be accomplished bit by bit. Saving the population of the state from genetic decline was the kind of emergency that required widespread and radical state action.

A condensed version of Anderson's report was integrated into the Committee on the Human Factor's chapter on "The People" in *Rural Vermont*. Whoever had the final editing of that text assured its readers that the plan to remove the people living in the hill towns was of course "for their own good and for the future well-being of their children."[31] But this version added a blunt, even brutal phrase: people should be removed "even though they would probably not be happy anywhere else."[32] Lemuel Peet, the author of the thirteen-town land utilization study, had been reluctant to suggest relocation for hill farmers, fearing that the people might prove to be unfit for any other environment and be harmed by removal.³³ This author, in contrast, seemed to be hoping that they *would* be harmed by it: the gravity of the threat to the state required that these people be relocated to places where, "in the competitive atmosphere of 'getting,'" they would encounter more hardship rather than less: places where "they would have to make use of any latent capacities" they had in order to survive.³⁴

In this way, without any explicit recognition of the change, the agenda of the Country Life Commission was turned upside down. The problem of the hill farms was no longer how to stabilize their populations—although that had originally been the reason for the creation of the Commission and the goal of the Eugenics Survey. It was now a question of how to "permit" (in the language of the Forestry committee) or to "encourage" (in Anderson's language) the farmers of the hills to do precisely what farm advocates for decades had *feared* they would do: abandon their farms and go elsewhere.

Gathering the Data

It was quite a sweeping analysis, and the eugenics team collected a great deal of data to back it up. Anderson, Rome, and Choate spent months in each of the three towns, conducting between 150 and 180 lengthy interviews in each, sometimes returning to a household more than once. Their notes suggest they drank many, many cups of tea and coffee in farmhouse kitchens. To each visit, they brought an elaborate set of printed forms, with designated spaces for answering dozens of questions. Just devising the questionnaire must have been a complicated process.³⁵ Filling them all out was an extensive undertaking.

In one section, there were spaces provided to rank the condition of the farm, the house, and the "social status" of the family. At the top of the first page, they were to note the religious and ethnic backgrounds of the residents. A chart provided space to enter the birthplaces and educational attainments of the household's husband and wife, their children, and their parents. In another section, the investigators were to record a variety of economic data: the number of livestock the household cared for; what crops they grew; the household income and its main source; whether their property was mortgaged and if so, for how much; what the land was worth; how many acres they had; and even their relationship to the previous owner of the land. In yet another section, the interviewers recorded responses to the open-ended question of why the members of the household had not (yet) left town. (In a different set of documents, the eugenicists attempted to account for everyone who *had* left town, where they had gone, when, and why.)

If that were not enough, the interviewers also had space to write their own comments. Sometimes they used that space to clarify the other responses—calculating gross and net income, for example—but more often they ranged freely, describing the residents' physical appearance, their children's apparent intelligence, their furnishings, the state of their health, and even their willingness to talk to the investigator. Back at the office, they tabulated the data that was quantifiable.

The investigators were also forced to do more than a little emotional work—perhaps more in Jamaica than in their earlier surveys of Waitsfield and Cornwall. They often recorded their own frustration or annoyance in the field notes, and they sometimes reflected on their own efforts to understand and come to terms with what they heard. As it happened, each of the three investigators came from a background that would necessarily complicate her reactions to Jamaica and its problems. Anderson herself had been born to immigrant parents (her father a teamster) in what was then the small town of Selkirk, part of a large, tight-knit Icelandic settlement on the plains of Manitoba. Her remarkable journey from there to Winnipeg and the University of Manitoba—and from Manitoba to New York City and Columbia University's School of Social Work—must surely have shaped her perspective on the question of rural outmigration.[36]

Anna Rome had a different story. Like Anderson, she was the daughter of immigrants: her Jewish parents came from Lithuania to Burlington, Vermont, only five years before she was born. Like Anderson, Rome made a long journey, although hers was not measurable in miles. Her high school degree and training in stenography earned her an office job with the Eugenics Survey, taking her from the "Little Jerusalem" neighborhood of Burlington's Old North End uphill to the imposing edifices of the university.

In contrast, Marjorie Choate was the daughter of well-to-do native-born parents and a graduate of Syracuse University. She had no migrant story—but her grandfather did. Like many boys of his generation, he had left his farm home and migrated to a city in search of work. Elliott Choate ended up in the industrial city of Fitchburg, Massachusetts, where he worked his way up to become a successful businessman. (In 1900, he told the census taker his occupation was "capitalist.") It is hardly surprising that he grew up on a farm: the surprise is that the farm he grew up on happened to be in the town of Jamaica, Vermont.[37] Of course, it is possible that Marjorie Choate had never visited Jamaica before her employment with the Eugenics Survey, but she must surely have heard stories about it in her earlier years.

All three investigators thus had their own reasons for pondering questions about migration, identity, and mobility. And that is to say nothing of their shared experience with the difficulties women faced in the workplace at all levels. Landing the new and relatively prestigious job of "eugenics investigator" was a piece of good fortune for all three. (In fact, those jobs were mostly filled by women, offering important new opportunities especially to those who, like Anderson, had earned advanced degrees.[38]) The attitudes and opinions they brought with them to Jamaica often informed what they wrote in the field notes.

The Conclusion Upon Which the Data Were Based

Still, after three months and 158 interviews, Elin Anderson's final account of Jamaica would be entirely consistent with William Frazier's initial comment about the town's "lack of initiative." Just as Frazier had predicted, the team found that the residents of Jamaica were "not driven by the will 'to get on,' as are the people on the other side of the mountain" (in the bustling commercial center of Manchester).[39] And the field notes reveal the process by which Anderson came to this almost, but not quite foregone conclusion.

To be sure, there was some support for Anderson's analysis of Jamaica in the data her team collected. A handful of outspoken residents of Jamaica actually concurred with Frazier's judgment about the community's lack of initiative. In 11 of the 158 interviews, those residents asserted—often in more aggressive terms than those used by the investigators themselves—that their neighbors lacked ambition or even that they were lazy. One woman "remarked about the laziness of the farmers in J, said they could make a success at farming if they only had the ambition." She told Anna Rome that her own family was "able to make a fine living off their farm and there is no reason why others shouldn't."[40] A second resident told Marjorie Choate that "the present generations in J show a decided lack of enterprise." This man shared the common belief that "there is no question but that the most ambitious have left J."[41]

In other cases, informants endorsed Anderson's idea that there was something peculiar about the tranquility of Jamaica's atmosphere. One Christian minister described Jamaica residents' lack of ambition as a kind of infectious disease endemic to the town, imploring Anderson to assure him that he would not catch it from his neighbors.[42] Another man, a physician who had moved to Jamaica from a different part of Vermont, offered an almost anthropological analysis, describing the town in terms of its primitive culture. He told Anderson that Jamaicans were "typical of the people anywhere in Vermont who live in a mountainous section such as this. There is infinite contentment and no worry about the future."[43]

While they were working in Jamaica, the three investigators boarded with a family in the village.[44] The notes Elin Anderson took about her conversations with the landlady's husband are incomplete and appear in the records out of order, unattached to the other interviews. This man's viewpoint seems to have piqued Anderson's interest and helped to shape her own interpretation of the situation in town. He suggested that the problem with Jamaica farmers was not their stony soil or hilly fields. Despite the "marginal land" label, he insisted that "it isn't that those fellows ain't got good land." It was that the farmers "don't have either enough money nor planfulness to make it pay."[45]

The idea that the farmers did not have enough money to farm was repeated by several informants, and it raises questions about the nature of farming in Jamaica—matters that will be discussed in chapter 4. But the assertion that most interested Anderson was the one about "planfulness." In her notes, Anderson debated with herself about the landlord's words. She was clearly reluctant to believe there was anything seriously wrong with the genetic inheritance of the people she was interviewing: "The class of people [in Jamaica] seems to be as good as that in any other Vermont town," she wrote. But she wasn't quite ready to accept the landlord's explanation, either, although it corresponded with what she had been led to expect. "It seems a little difficult to believe that they so lack in planfulness [the landlord's word], initiative [Frazier's word]," or "willingness to assume responsibility [Anderson's own phrase]." Could it be, she wondered, that they did not appreciate "the independence that a farm offers"? In these undated notes, Anderson arrived at a tentative diagnosis: "so far there seems to be a type of person in the town—perfectly content with things as they are."[46]

Perfect Contentment

That phrase—"perfectly content with things as they are"—would be repeated again and again in the investigators' notes, as both summary and judgment. It was the in-house phrase they used to identify the characteristics that would lead to Anderson's diagnosis of the problem of Jamaica. Despite the mountains of information recorded in these interviews about the residents' education, farming practices, and household finances, this matter would claim an outsize portion of the investigators' interest. In over a third of the interviews a question was raised about the resident's work ethic, ambition, or level of "contentment."[47] The discourse appears to have been shaped primarily by Anderson herself, who recorded half of those interviews.[48] The phrase "perfectly contented with things as they are" was a favorite of hers: Rome and sometimes Choate described some residents with the more traditional epithets "shiftless" or "lazy," but Anderson avoided those harsher words almost completely. Choate sometimes used the word "content" in a neutral way: an elderly couple are "thoroughly content and enjoying life," a pair of brothers are "content to live simply." But for Anderson, "contentment" almost always carried the implied judgment that the person ought to have been very much *discontented*: with primitive living conditions, an insecure or low income, or the lack of opportunity for advancement. In her usage, to be "perfectly contented" with such a state of things was not so much a moral flaw (like laziness) as it was a kind of mental or perhaps psychological deficiency.

All three investigators seemed to agree that ambition was an entirely admirable character trait. In nine interviews, a resident was described as having the right amount of ambition, but usually in comparison to someone else in the family who did not: one woman, for example, was described as the "really ambitious" member of the family, another as "ambitious for her children's education." In thirteen interviews, a resident was described as not ambitious enough: a daughter whose desire to attend school was "not quite strong enough," or a son who "never had the drive to leave."

It is worth recalling that not very far in the past, farmers were routinely praised for their lack of ambition. Agricultural experts who had weighed in on the problems of the New England hill town in the late nineteenth century often praised New England farmers precisely because they chose to stay at home rather than exercise their ambition and go out West or to the cities to compete for fame and fortune. In the Board of Agriculture's annual report for 1875, for example, a commentator deplored the fact that the recent Civil War had led many rural men "to run wild with hazardous and unscrupulous haste into any scheme that promised to enrich themselves." In contrast, it was the Board's duty to "show to the young man about to choose a calling" that farming was "the freest from care and anxiety, wasting weariness and nervous perplexity of any engaged in by man."[49] "Ambitionlessness" had in fact been a feature of the idealized farmer in agrarian fantasies for centuries: orators habitually praised the simple man who realized that a "competence" was a better goal than wealth or fame.

In the twentieth century, Vermont's agricultural experts were less sure about the proper role of ambition in the farmer's life. Was the ideal farm still primarily a stable and secure home where the family found safety, food, and shelter (as Professor Peach described it in his 1929 committee report to the Country Life Commission)? Or should the farm be a modern business operation taking risks and competing for profits as city business owners did (and as Congressman Brigham implied in his remarks at that same gathering)? Perhaps without thinking too much about it, Anderson shared the assumptions of most of her scientifically educated progressive peers. As the phrase went, farming should be regarded as a "business," not a "way of life." If the profit and loss figures didn't work out, the business would fail, and the worker would move on.

Anderson was not without sympathy for the people she interviewed, but she seems frequently to have been exasperated with them. She was particularly frustrated with people who expressed a preference for strong family ties over good job prospects elsewhere, and most of all with young people who would not leave home. Perhaps that is not surprising coming from someone who traveled so very far in pursuit of her own professional opportunity. Describing one family, Anderson characterized the household as "content to go their way" and the husband as "genial and content," a phrase that in context was clearly intended as a reproach: the fifteen-year-old son in this family had turned down his teacher's suggestion that he go to high school in an adjoining town. (Only some Vermont towns supported high schools, although all students could attend. If one's hometown did not have a high school, taxes would pay the tuition for the student to attend elsewhere, but that would still leave transportation and perhaps boarding expenses for the family to cover.) This boy was "blind to possibilities," Anderson noted, and "seems to accept that this is the way of life."[50]

Another twenty-year-old son was still living with his family when Anderson visited. "His father thinks that maybe he is staying to be near him," she wrote, but she herself thought that would be "a pretty poor excuse."[51] In another family, Anna Rome reported "a daughter, aged 19, and a son, aged 18," who were "staying at home and doing nothing, except for helping around the farm and house." (Presumably by "doing nothing," she meant nothing for wages: after all, "helping around the farm and house" would seem to be describe the normal work of men and women on farms.) Rome found it unaccountable that the mother seemed "perfectly satisfied with things as they are." The mother acknowledged that her children would probably have to leave to find paying work eventually, but she was not looking forward to that day. Rome commented: "Doesn't seem to mind their staying home and not working, even though they just manage to get along."[52] Anderson encountered that same baffling attitude in yet another household, in which the mother actually boasted that "her children didn't go away."[53]

In one case, Anna Rome pointedly asked a resident if her sixteen-year-old son was planning to leave town and seek work elsewhere, but the woman "said that she didn't think so for a while as he was so young and they didn't want him away from home." She was also reluctant to have her son do heavy work around town because of his youth. Rome noted that she herself thought the boy seemed "quite grown up and capable of working." Reinforcing her own judgment, she recorded the opinion of "a neighbor," who told her the boy was "too lazy" to work.[54] It is difficult to know what to make of Rome's response to all this. She herself was just twenty years old at the time. She had a job, to be sure, but she was still living with her own mother in Burlington.

The team's concern with children who would not leave home is curious, not only because of their personal investments in the question, but because a large part of the original mission of the Commission on Country Life had been to prevent young people from leaving farms and rural communities. And to compound the confusion, in comparison with the other two towns that were part of the Migration Study, Jamaica was actually the town which had the greatest success in keeping young people at home. As Anderson reported, "In hilly Sylvania, more than in the other two towns, grown-up sons and daughters have remained."[55] In theory, that should have been perceived as a sign of a healthy community, but that is not how the interviewers saw it. In their view, Jamaica's home-keeping daughters and sons were part of the problem. (And it was specifically the children of Jamaicans—and not those in the other two towns—who were described in those negative terms. With very few exceptions, the field notes from Waitsfield and Cornwall simply noted without remark: "Boys very content on farm."[56]) The problem seems to have been something about their *reasons* for staying home.

Reasons to Stay

What was it that kept some Jamaicans at home, where it was "always afternoon" and jobs were scarce? There were 279 separate responses recorded in the section "Reasons for Coming or Staying."[57] The interviewers paid a great deal of attention to that section, devising a complex set of categories to fit the responses. A penciled tally in the Eugenics Survey files shows the effort that went into ticking off, adding up, and averaging the numbers.[58] And Anderson used the "Reasons for Coming or Staying" responses extensively in her published reports. One finding proved to be particularly important to her overall analysis of Jamaica's problems. According to the team's tabulations, most people who *left* Jamaica reported an "economic" reason for doing so (as did those who moved into town). Anderson interpreted those comments as expressions of initiative. In contrast, according to their count, very few people gave an economic reason for *staying* in town.[59] Most commonly, Anderson reported, the group who remained at home explained their reasons with what she perceived to be a lackadaisical expression, something like "Always lived in town and never thought of going away." Those kinds of phrases confirmed her suspicion that Jamaicans lacked initiative. Yet those kinds of responses could easily have meant something else altogether (see Table 5).

A total of 35 percent of the interviewees in Jamaica offered reasons for coming to or staying in town that revolved around either family or community connections. Twelve percent of these gave reasons related to family. Twenty-three percent explained that it was the town or community that drew them there or kept them there. Some of these statements were the very ones Anderson interpreted as evidence of lack of ambition. Ten percent of all Jamaica responders, for example, said they stayed in town because they liked where they lived. Six percent of the responders simply said, "it's home." Perhaps they did not think such a sentiment required explanation. At any rate, it requires a significant leap of faith to interpret a phrase like "it's home" as an expression of any character trait at all, other than perhaps an excess of Yankee brevity.

Indeed, in order to interpret Jamaicans' responses as evidence of their lack of initiative, the team had to ignore one critical fact: the two "better" towns in the Migration Study responded in very similar numbers with phrases like "it's home" and "always lived here." (They did not appear to have liked their hometowns as much as Jamaicans did; Cornwall residents seemed to have liked their farms more.[60]) A little over a fifth of responders in each of the three towns gave a reason like that. Cornwall and Waitsfield also mentioned family at a rate a little higher than Jamaica's, not lower as one would expect in more ambitious towns (more about these family responses in chapter 6).

Moreover, because this point about lack of initiative was so central to her analysis, Anderson made a small but revealing error. As she reported, the young people of Waitsfield and Cornwall typically used positive and active phrases like "wanted to better myself" to describe their choices to leave town. Jamaicans, in contrast, often used what Anderson considered a more passive phrase: they were "no hand to go out and find themselves a job."[61] But Anderson had it exactly backward, at least if the team's field notes are accurate. That expression appears only once in all the field notes, and it was used by a resident of the thriving town of Waitsfield, not of struggling Jamaica.[62]

A second problem with Anderson's reading of the data has to do with those "economic" reasons for staying. By my count, nearly a third of Jamaica residents cited some kind of economic reason for remaining in town—just about the same as Waitsfield and significantly higher than Cornwall. That category, to be sure, included all kinds of considerations other than personal ambition. Some were expressed in positive terms: *I got a job in town; I've been doing well here.* Others were less than ringing endorsements of the economic climate in Jamaica, but nevertheless referred to important economic considerations: *things are cheaper here; I lost my job elsewhere; a person can make "as good a living as any" here.* Sixteen comments referred specifically to an opportunity offered by a family member: pooling resources with a brother or sister; inheriting a farm or a business; buying or renting a relative's farm at a good price. These were clearly "economic" reasons, but they also suggest something about the role of family and community ties in retaining residents—a subject Anderson did not investigate.

And family was not the only subject that was left untouched by the team's analysis. Perhaps most striking is that the interviewees'"non-economic" reasons for staying in town were far more complex and diverse than Anderson's analysis suggested. 14 percent of Jamaica residents offered a reason related to the physical place itself. No one brought up Jamaica's dramatic mountain scenery as a reason for living there (although perhaps that was part of what some residents had in mind when they said that they "liked the place"). But some said they preferred country living to city life; others preferred farming to other types of work; and a few said they wanted to live in New England, or Vermont, or just "this part of the country."[63] One couple reported, for example, that they had traveled by automobile up the West River valley in 1918 and had decided that they liked Vermont best "of all the states they had been in." That couple "particularly liked this location in Jamaica, so bought the place."[64]

The team's analysis of the Jamaica responses may thus not be altogether false, but it is deeply misleading. That becomes even more obvious in the lengthier responses. Most of the responses recorded by the interviewers were in the form of brief phrases—*"likes farming best of all," "very glad to be away from the city"*—but in a few cases, the interviewer recorded a more complete account that spelled out in more detail what brought them to or kept them in Jamaica. One young man, for example, told Anna Rome that he had "always wanted a farm and so he is glad of the chance to farm and expects to make his entire living off the farm."[65] A second man explained to Marjorie Choate that he believed "J and Windham are coming back as far as farming is concerned."[66]

88 CHAPTER 3

Table 5. Jamaicans' Reasons for Staying in Town

		JAMAICA	WAITSFIELD	CORNWALL
	TOTAL # RESPONSES	279	274	183*
	Always lived here	10	13	10
	Never considered leaving town	9	2	6
	It's home	17	17	5
	Likes town	27	18	4
	Always lived on farm	0	1	7
	Never considered not farming	0	4	9
HOMETOWN		63 (23%)	55 (20%)	41 (22%)
	Likes region	7	1	1
	Prefers country to city	11	5	3
	Wanted to be "on a farm" (not in town)	6	5	0
	Likes farming	9	37	26
	Likes this farm	3	4	2
	Doesn't like other work	3	2	6
PLACE		39 (14%)	54 (20%)	38 (20%)
	As good as any	5	0	0
	Cheaper here	3	1	1
	Looking for work	6	10	10
	Not fit for other work	2	0	5
	Doing well here	13	5	0
	Economic opportunity/job	33	40	14
	Lost job/farm elsewhere	8	6	5
	Family opportunity	16	20	8

ECONOMIC		86 (31%)	82 (30%)	43 (23%)
	Family obligations	10	16	17
	Family ties	19	11	11
	Marriage	4	8	2
	"Fell to lot"/ seemed natural	0	11	0
FAMILY		33 (12%)	46 (17%)	30 (16%)
HEALTH		23 (8%)	7 (3%)	12 (6%)
RETIREMENT		12 (4%)	7 (3%)	4 (2%)**

*Cornwall numbers are lower because fewer people answered this question.
**Percentages do not add up to 100% because of miscellaneous individual responses.

A third man told Anderson that he was worried about the worsening economic crisis that fall. He "doesn't know how some people will live over the winter," she recorded. "Men on most backwood farms are better off than people in the villages."[67] A woman agreed with him, but for a different set of reasons: "people who live in J are infinitely better off than those who have left to go to the city." That woman had cultural opportunities in mind when she said that: "nowadays with the good programs that are available on the radio and the magazines . . . and with the traveling libraries," a person in the country "has all the advantages of the city dweller."[68]

One young farm worker hailed from the industrial city of Fall River, Massachusetts. By chance he was interviewed by Marjorie Choate. The farm worker told her that "he like[d] Vermont and farm work very much in comparison to city jobs and hope[d] to be able to get money enough to have a farm of his own some day." Although she judged the family to be no more than "respectable laboring class," Choate labeled the young man "enthusiastic and likeable." Perhaps his response reminded her that her own grandfather had made the opposite journey in his life, leaving a Jamaica farm to "get money enough" to become a successful businessman in another Massachusetts industrial city.[69]

The most complex set of reasons for staying in Jamaica was given by a married couple. Anderson noted that the Armstrongs "wanted to get back out of the way [her quotes]." They "didn't like the noise of cities—preferred country." They had not been able to buy the husband's childhood home in nearby Stratton, so they had bought the farm "nearest to home possible that was this place." Mrs. Armstrong added that she thought it best "to be out on a farm to bring up children," and reiterated that she and her husband 'want to get back out of the way.'" Both of them, she repeated, "had had enough of cities." Anderson was unable to resist the temptation to add a skeptical comment: "Actually neither one of them have lived much in a city."[70]

The Armstrongs had one more strong reason for preferring to live in the country. In 8 percent of all answers, ill health in the family was given as a chief reason for being in Jamaica. Those responses were usually connected, implicitly or explicitly, to Jamaica's location. Because of the contemporary view that mountain air was therapeutic, particularly for respiratory ailments, the higher elevation of Jamaica added to its appeal. (In Waitsfield and Cornwall, at lower elevations, health was less often mentioned.) One man reported that he had returned to Jamaica "because health failed him while in the paper mills and the doctor said he would have to be out in the country."[71] Another explained that his asthma was too bad to live in his old home in New Hampshire, and a woman told Anna Rome that her children "were not well [in Lisbon, New Hampshire] because of smoke from factories."[72] For still another, Choate noted: "She came to Vermont in 1918 account [husband's] health (bronchial)."[73]

As to the couple who wanted to get "back out of the way," Mr. Armstrong was a World War I veteran who had come home with shrapnel in his lungs and mustard gas injuries as well. After the war he had opened a garage in Westmoreland, New Hampshire, but within a few years, he and his wife moved to Jamaica, perhaps seeking out the mountain air as a treatment. (Westmoreland is on relatively low land beside the Connecticut River.) On top of her husband's problems, Mrs. Armstrong reported that "a Christian Science woman" had suggested that she herself "get closer to nature" as a treatment for her chronic illnesses. "Now since she has come to J she feels perfectly well. Her dropsy is cured and the leakage has stopped."[74]

The three towns in the Migration Study were roughly equal in the numbers of residents who responded with phrases like "it's home" and "always lived here." But Jamaica residents were more likely to say that they "liked" the town. They were also more likely to say that they "preferred the country to the city." Jamaicans were more likely to be retirees; and many more residents reported that they came to Jamaica or stayed there because of their health. Taken together, these numbers begin to paint a portrait of Jamaica quite different from the one Anderson described in her report.

Leaving but Returning

There was one more set of facts the investigators missed in their data, probably because they simply were not expecting it. A common joke about hill towns was that their residents led such narrowly constrained lives that they had literally never traveled beyond the boundaries of their town. Anderson repeated a joke of this kind she picked up from a resident: Two Jamaica men were arguing, and as an insult, "one told the other that he ... had never been out of the town anyway. The other replied, 'I have too. I was over to Stratton one day.'" As Anderson explained, "Stratton is a heavily forested mountain nearby." (That was not quite correct: Stratton at that time was still a functioning town, although in a few years a ski resort built on that mountain would transform the place.)

Table 6. Migration of Native-Born Jamaicans

Total Native-Born	103
Never left town	27 (26%)
School or war only	6 (6%)
To/from adjacent towns only	20 (19%)
To/from farther places	50 (49%)

According to the data gathered by Anderson and her team, however, only a small number of Jamaica residents could properly be described as having "never left Jamaica (see Table 6)." On the contrary, nearly half of the 103 heads of household born in Jamaica and adjacent towns had moved beyond the boundaries of Jamaica and its surrounding towns to live elsewhere, often for many years, before returning to Jamaica.[75] Another twenty heads of household had moved around among the contiguous surrounding towns. Altogether, over two-thirds of native-born Jamaican heads of household had at least some experience with living in other places before returning to town. Anderson labeled the residents who left town and then returned "repeaters." Out of the three towns the eugenicists studied, Jamaica reported the largest number of those who left town and later came back.

Local migration back and forth among small towns like Jamaica and its neighbors is not surprising. It was not only normal, but necessary for the social and economic health of the community. Jamaica is a town of widely scattered villages and neighborhoods, each cut off from the others by mountains and rivers. For some people, the surrounding village centers in Wardsboro, Bondville, South Londonderry, South Windham, and West Townshend were more accessible than Jamaica's own center villages. Men and women born in Jamaica married partners from nearby towns; they moved to farms on either side of town lines. Men working in sawmills or in the woods followed the jobs from Jamaica to Windham to Stratton and to Jamaica again. Women crossed town lines to work as housekeepers or nurses for employers in other towns.

Residents who owned no real estate and relied on day labor for their support were particularly likely to move from one nearby village to another as jobs at farms, sawmills or lumber camps became available or dried up. One man reported to Marjorie Choate that "since his marriage he has lived in W. Brattleboro, Dummerston, Putney and J. Worked as a laborer and went wherever he could find work."[76] Another reported to Anderson that he "went to Brattleboro for about 2 years and drove a truck there, but liked home best—so returned to farm and work out. His wife wants him to get started on a farm, and they are planning to buy one shortly in Londonderry."[77] It was especially common to move upon marriage, when either the wife or the husband might have access to a farm or work opportunity across a town boundary. Anna Rome reported about one young man that he had been born in Winhall, worked as a hired man there, "did odd jobs," and "at one time was in Chester and Londonderry working on the telephone lines for short periods"—all before he married a Jamaica woman and moved to town.[78]

But the largest group of migrants from town traveled much farther than the surrounding neighborhoods. Fully half of the native-born heads of household traveled out of the area, often far away and for years at a time before returning. (And this list almost certainly understates the mobility of the community, because women discussed their travels only infrequently, either because they were not asked or because they chose not to. With a few exceptions, it was the migrations of male heads of household that were recorded.) The great majority had traveled to Massachusetts, New Hampshire, or other (non-contiguous) parts of Vermont, but residents also reported traveling to Connecticut, New York, Maryland, and Pennsylvania—and also as far afield as Texas, Iowa, Illinois, Florida, the Dakotas, and "the West."

One man reported to Choate that he had been born on a farm very near where he currently lived. He had moved around a lot before returning to his childhood home: "As there was nothing that he cared to do in J," he had left as a teenager, moved to Massachusetts, and "from there to the Pacific Coast," working as a carpenter. He returned to Jamaica in his thirties. The notes explained: "Left to get work that interested him; returned to be near his parents."[79] An elderly woman described a similarly winding path back home. She had first left Jamaica to work as a domestic in her uncle's family in Boston. There she married a minister with whom she lived in Massachusetts for many years. After his death she spent fifteen years alternating between Jamaica and her son's home in Connecticut. In her sixties, she re-committed to her hometown by marrying another Jamaica native.[80]

Newcomers

All these data clearly point to the fact that native-born Jamaicans were a surprisingly mobile group, and evidently not prevented from traveling by any lack of "initiative." And the data reveal something else that must have caused trouble for the investigators. For a town in such evident demographic decline, there were a surprising number of new folks around. Most people in Jamaica were native-born, to be sure: two-thirds of the town's residents in 1930 had been born in Jamaica or one of its contiguous towns (or had come to town as children with their parents). But a full third of the heads of household in 1930 had been born somewhere beyond those nearby towns: in other parts of Vermont; other parts of New England; the northeast, or even farther (see Table 7). Some came to Jamaica looking for jobs or farms; some came to retire; some visited first as vacationers and then settled down. Many were connected to Jamaica by family ties: a wife's relatives, a daughter who married into a Jamaica family, or a family member who had moved there.

Table 7. Origins of Migrants to Jamaica

Total Non-Native Resident Heads of Households	54
Vermont (non-adjacent towns)	22 (40%)
Other Northern New England	3 (6%)
Southern New England	12 (22%)
Other Northeast	13 (24%)
Other (Arkansas, Great Lakes, Canada)	4 (7%)

Neither the mobility of the "repeater" Jamaicans nor the new migration to town suited the theories of the investigators. It was difficult enough to explain what brought people back home to remote, declining Jamaica. It was even more difficult to explain that so many people from other places found the town appealing enough to move there. If, as the investigators were concluding, there was something in the environment that sapped the initiative of its native-born people, perhaps people who had grown up elsewhere could not (yet) feel it. Choate wrote about one family, for example, "One of the more enterprising families in J, possibly because they are newcomers."[81]

Or perhaps there was something already amiss with the new migrants that attracted them to the town. As Anderson explained, the "immigration into Sylvania appears to be of a different character from that into Pomona and Beaufield [the pseudonyms chosen by the Eugenics Survey for Waitsfield and Cornwall]." The other two towns had their own problems: Waitsfield attracted "somewhat inferior stock," (a reference to hired men brought in to work for a handful of large landowners); Cornwall had attracted "foreign stock," mostly from Quebec. In Jamaica, the problem was different: the people coming in were "able but maladjusted." Although many were "well-educated people, a few even talented," Anderson judged that they were "running away from life." Jamaica simply "provided them with the means for escape." She gave examples of several such people in her published account: a young man who had come home to restore his father's farm; an elderly Finnish woman who lived alone in the woods; and the couple who had liked the area so much when they drove up the West River valley. The wife had an exotic background: a graduate of Vassar College who worked for the Red Cross during the war, she had traveled throughout Europe and the United States and had even survived the Titanic disaster. In her notes, Anderson diagnosed her with a "grandiose mental disorder."[82]

Whether "maladjusted" or not, it was undeniable that an increasing number of people were moving to Jamaica in the 1920s and that native-born Jamaicans were in motion as well. Fewer than one fifth of Jamaica heads of household in 1930 had lived in town their whole lives without any interruption. Two-thirds (including native-born travelers and incoming immigrants) had traveled beyond Jamaica's neighboring towns, either before or after living in Jamaica. Even those who never left town must have known numerous relatives and neighbors who had gone to Hinsdale, New Hampshire, for factory work, to France to fight the Germans, to New York City, or out West to see what it was like. Living in Jamaica was no longer, if it had ever been, simply the unexamined result of having been born there. It was now a choice.

Jamaicans told the investigators many complex stories of emigration and return, which the three duly recorded. One man told Anderson that he had left Jamaica to work in machine shops in several different nearby towns, returning home frequently. Most recently "he went to New York City and worked with the American Can Company but didn't like it and came back to J, even though he didn't have any work here."[83] Another resident told Anderson that he had "been in the Dakotas and far west, but prefers back home here."[84] A third told Anna Rome a long tale of his on-and-off relationship with Jamaica. Mr. Dewey had "spent his childhood mostly in Brooklyn, NY," but had come "to J when about 17 to board with relatives because he liked it here better than in the city." In 1894, "aged 23, he returned to Brooklyn, where he was conductor on a street car" and "enlisted in the Spanish American War." After the war ended "he was again a street car conductor," and finally in 1900 "he came to J to stay, preferring country life to city life." In a string of "reasons for staying" that went completely unmentioned in Anderson's summary, Mr. Dewey said he "liked it better than the city," and that he preferred "country to city life." And finally, as Rome summarized with no comment: "Says there is more freedom here."[85] (That remark will be discussed at greater length in chapter 6.)

Of course, some Jamaica residents never got far beyond their own hillsides. Some ran up against hard economic limits that kept them from pursuing the educational or job opportunities they wanted: one man told Choate that he was not "fitted" for any other work but what he could get in Jamaica, and another explained to her that he did not have "enough education" to get work elsewhere. A third explained that he would have liked to buy a farm in another area, but that good land was too expensive, and he had been forced to make do with a farm in Jamaica. Nevertheless, the field notes make it clear that most Jamaicans at some point or another had an opportunity to choose a different place and way of life. For them, the decision to go or not to go was not based on passive acceptance of the familiar or on lack of "initiative." It was based on considerations that the investigators mostly overlooked.

Jamaica was a town that had long been experiencing both population decline and dwindling resources. It was indeed a place that many sons and daughters were leaving behind. Yet it was also a town where new migrants showed up each year. And it was a town whose residents—most of whom had traveled around the country enough to know something about their alternatives—expressed a significant degree of satisfaction with their lives. Ultimately it was this conundrum that forced Anderson to come up with her complicated theory about the unhealthy contentment of the residents. Given the constraints of her assumptions and the requirements of her job, there were some answers she could not afford to hear, no matter how carefully she recorded them. That was the paradox at the heart of the investigation.

What Mrs. Tower Said

One interview conducted by Elin Anderson exemplifies how the interviewers managed to record so much while comprehending so little. On that day, the lady of the house, visibly pregnant, did not get up when Anderson came to the door, but stayed seated in her rocking chair and reached over to the door to let the visitor in. As Anderson recorded, Mrs. Tower was polite but appeared uninterested in answering questions. Still seated, she "rocked comfortably and answered the questions laconically"—"yes," "no," "I don't know." Anderson allowed her frustration to show in her notes: "Even any switching around to other topics failed to make her more conversant." Perhaps hoping to provoke some response from her on that late October afternoon, Anderson asked the woman if she "didn't dread winter coming." But even that question failed to make a change in the woman's placid demeanor. She simply replied, "No don't bother me any. I like it all right. Long as I've got enough food and enough to keep me warm, I'm alright." "And," Anderson recorded, "she kept on rocking gently."[86]

Anderson's question may have been driven by frustration, but it probably also reflected real concern about this family in the bleak autumn of 1930. Earlier that year, it might have been possible to imagine that the bad economic news would be confined to "city unemployment," as one Cornwall farmer had described it back in May. By October, it had surely become clear that rural people would also feel the impact of the deepening economic crisis. Mr. Tower earned $3 a day at a nearby sawmill when the mill was operating, but that had not been often lately. Knowing that they had two young children and another on the way, it is no wonder Anderson asked Mrs. Tower whether she dreaded the coming winter.

Yet Anderson did not ask the follow-up question that now seems obvious: Why *wasn't* she dreading the winter to come? How was she able to maintain that imperturbable calm? A reader today might speculate that Mrs. Tower was not being strictly honest with Anderson. Jamaica residents were surprisingly generous about sharing information with the investigators, but there were doubtless things they wished to keep private. Perhaps she was worried about the winter but unwilling to discuss it with her out-of-town guest. Mrs. Tower was a member of one of the town's "old families," as Anderson reported; her father was a "good farmer" who lived "right near." Perhaps she was relying on his help if things got bad. Or perhaps she did not want to advertise her husband's failure to provide. Anderson did not ask.

But she did record a kind of answer. In their zeal for their work, the interviewers often recorded more than they knew, including a great deal that did not fit easily into their own explanatory frameworks. And in the section of the questionnaire describing the operations of the farm, Anderson did just that, jotting down what amounted to Mrs. Tower's answer to that un-asked question. That year, the family had grown "just enough potatoes for ourselves;" they were raising "only a few chickens—just enough for our own use," and kept "two cows—just enough milk for ourselves." Anderson seems to have heard Mrs. Tower's statements as a kind of apology: "only" a few chickens, "just enough" potatoes and milk. She put quotation marks around Mrs. Tower's explanation that they "didn't have time" that year to put in any more than a small garden. (The quotation marks suggest that Anderson did not believe that excuse; she had heard other residents say that some Jamaicans were too improvident or "planless" to prepare for the spring planting season.) But a closer look at Mrs. Tower's words and even body language (as reported by Anderson herself) suggests that this woman was not apologizing at all. She may even have been boasting a little. If her husband earned little that winter, there would be hardship in her house, no doubt, but there would be food on the table and a fire to keep them warm. Anderson, caught up in another narrative, was apparently unprepared to hear the hint of satisfaction in such a statement, but a listener accustomed to scarcity—perhaps one of Mrs. Tower's own neighbors—might have caught her tone.

In fact, Mrs. Tower conveyed more than her family's subsistence strategy to Anderson. Without being explicit, she told her a little about her values. Anderson noted, for example, that the Towers' oldest daughter, aged eight, typically walked a mile to school and back "even in winter." Anderson commented disapprovingly that the "Mother doesn't think that any task." Mrs. Tower evidently valued self-reliance and hardiness. As she told Anderson, she had made the very same walk to school "when she was a girl and thought nothing of it." (And she did mean the "very same walk." She had grown up on the farm next door, so she knew exactly what her daughter's walk was like.) A second look at that other response—"Long as I've got enough food and enough to keep me warm I'm alright"—might also suggest that it was weightier than Anderson thought it was. Someone more self-dramatizing than Mrs. Tower might have said: I am a woman of modest tastes, contented with simple living. Making do with little, she might have reminded Anderson, was (or at least once was) a virtue highly prized on Vermont farms. Complaining about plain living was not.

As Anderson did in this interview, the investigators compiled an immense amount of financial, agricultural, and personal information about the people of Jamaica. Much of it, they did not fully comprehend. Yet, paradoxically, these interviews would offer many Jamaicans perhaps their single best chance to tell their own stories. Residents like Mrs. Tower seldom succeeded in getting the investigators to understand or approve their viewpoints, but they are at least well represented in the documents they left behind. The next chapters will explore what the Jamaicans told the eugenicists about their farms, their work, and their lives.

CHAPTER 4

What Farming Really Meant

In 1880, Jamaica resident Nat Price was eighteen years old and living on his father's farm. In 1930, when he spoke with Elin Anderson, he was the 68-year-old owner of that same farm. Dramatic cultural and economic forces had altered the landscape over those fifty years. Yet as he described things to Anderson, it did not seem like all that much had changed.[1] Back in 1880, his father's operation had been a typical hill farm, much like those described in chapter 1. The family had raised a wide variety of crops both for home use and for market, including potatoes, apples, beans, pork, and eggs; their cash crops were chiefly butter and maple sugar. By 1930, Price had switched from butter to milk, and from maple sugar to maple syrup (both more modern products), but he was still milking cows and tapping trees as his father had done. And like his father and his grandfather before him, Mr. Price still operated a small on-farm shop—"the same wheelwright shop of his grandfather—a much cluttered up little place," Anderson wrote. There were no longer any wooden wheels to mend, but he used the tools he had inherited to repair neighbors' furniture.

In fact, Nat Price's operation had almost certainly changed even less than the interviewers' data shows. They usually took note of the basic cash crops raised by residents, but they recorded very little about any kind of farm production that was not clearly intended for market: there were few references to eggs, garden vegetables, or firewood in their records. Statewide averages suggest that, in addition to milking cows, tapping trees, and repairing furniture, Price was likely still cutting his own firewood, and probably raising corn, beans, potatoes, and a pig or two for home consumption, if not to sell. Some aspects of his operation had certainly changed, but in some important ways, it was still his father's farm.

Trees Returning, People Departing

Still, Mr. Price was contending with dramatic changes in the land all around him: and, in his generation, the rate of that change was accelerating. The town lost almost a third of its population between 1900 and 1920, falling from 800 to an all-time low of 566 residents, where it would remain stable for another forty years. The mills and workshops that had crowded the central village in Price's youth had now mostly shut down. At its high point in the late nineteenth century, Jamaica had been home to almost all the usual small nineteenth-century businesses: harnesses and leather goods; boots and shoes; carriages and sleighs. Wooden products had dominated manufacturing in town; at one point, eighteen separate mills were making shingles, rakes, trays, tubs, chairs, coffins, and tennis rackets.[2] By 1930 almost all the small artisanal shops like Price's wheelwright shop were gone, and only four woodworking manufacturers were left.[3]

In farming, there was an even greater change. From 1860 to 1880 there had been only a small decline in the number of farms in Jamaica.[4] But between 1880 and 1900 the number of farms decreased by 17 percent. And between 1900 and 1930, the pace of change picked up dramatically, as Jamaica lost 41 percent of its remaining farms.[5] Nat Price had grown up in a town with over two hundred farms; as he reached old age, fewer than half those farms remained. Of course, a drop in the number of farms might not necessarily have meant a decline in the use of land. It might mean that some farmers were consolidating the land of others into their own holdings. And that is part of what was happening in Jamaica: between 1880 and 1930, the average size of farms in town grew from 120 acres to 156 acres, as farmers took over all or part of the land of neighbors who left. But it was not just that the land was changing hands: by 1930, the agricultural census recorded that nearly two-thirds of Jamaica farmland was now wooded.[6]

The Price farm was not immune to these sweeping changes. The 1890 probate record of Nat Price's father, for example, indicates that in addition to the "home farm," he had acquired a hundred-acre parcel described as the "hill farm," perhaps purchased from a neighbor who was no longer farming.[7] The details Elin Anderson recorded in 1930 about the Prices' "extra farm" suggest that it was well along in the process of reforestation. What remained of its open land was now used only for grazing: "they no longer cut" hay from those fields. (The thirteen-town study of the hill towns found that this shift from hay fields to

grazing was often a step on the way toward "abandonment" of a field to trees.⁸)
The home farm was changing, too. In 1880, it had been mostly open, with just 20
percent of its land in woods.⁹ Now, Anderson wrote that only "part" of Nat Price's
home farm was still "open near river." The rest of it, too, was returning to trees.

FIGURE 6. "Pasture scenes showing an intrusion of weeds and trees on the cleared land." This
photograph was taken by the thirteen-town research team and included in Peet's master's thesis.
In the foreground is a closely grazed enclosure, but beyond the stone wall, there are bushes and
weeds in the fields, and in the far distance the trees seem to be encroaching on the border, where
the land is increasingly "slopy." Source: Lemuel J. Peet, "Problems in Land Utilization in the Hill
Towns of Vermont, Based on a Study of Thirteen Towns in 1929" (master's thesis, University of
Vermont, 1930), 115. University of Vermont Archives, Silver Special Collections Library, University of Vermont.

Like their neighbors, the Prices were adapting. The "extra farm" acquired
back when the father was still alive appears to mark the beginning of a shift
happening everywhere in town—toward less intensive land use practices and
more reliance on the products of the re-grown woods. The Prices still raised
dairy cattle, but the herd grazed now on the old upland pastures of the second
farm, or perhaps among the maples: his maple "grove" was now located up on
that reforested second farm, too. The 1930 *Soil Reconnaissance Survey* explained
that it was common practice in the hills to use the maple groves for this purpose:
"In the higher sections many sugar maple groves grow, from which most of the
underbrush has been cleared, and the land is used for pasture."¹⁰ That may not
have been "best practice" if the Prices had been aiming to modernize their dairy
operation. (Modern milk cows would probably require higher quality forage.)

Perhaps it was a sign that the dairy herd was less central to their operation now than it had once been. Price informed Anderson that his most important moneymaker was now his maple syrup.

Still, it is not surprising that Mr. Price seemed more aware of continuity than of change. He still walked to the barn for milking every morning and evening, every day. He and his neighbors still used their land in familiar ways, grazing their cattle on stony uplands, plowing and hauling manure to their most level and fertile fields. And their seasonal routines changed little, if at all: they cut hay in summer, wood in winter; tapped maple trees in spring; dug potatoes and picked apples in fall. (In the late fall and early winter, they rested a little—a subject for chapter 6.) In these ways and many others, Jamaica farmers were literally doing as their parents and grandparents had done.

But of course, they were no longer living in the world of their parents and grandparents. Everywhere in the hills there were experts now, studying the problems confronting these farmers, eager to offer their advice about how to adapt. And at every turn there were obstacles—obstacles in the way of modernizing, and equally daunting obstacles to farming as their grandparents had farmed. For most, it would not be possible simply to modernize their operations; nor could they simply maintain old ways. The third option, of course, was to leave the farm—but that was not a trouble-free path, either (more about that in chapter 5). Confronted with these difficult choices, Jamaica farmers devised a set of strategies of their own, often completely at odds with what the experts advised.

Barriers to Modernization

By 1930, it was a popular assumption among many progressive Vermonters that the greatest barrier to success on the hill farms was the newly labeled "class 6 steep stony soil" of the uplands. But Jamaica farmers had been coping with those growing conditions for generations. From their standpoint, the soil was not the source of their predicament.

The eugenics interviewers themselves rarely made independent judgments about soil quality. (That was not their area of expertise, of course.) There was a section on the questionnaires labeled "Condition of Farm," but there the interviewers typically wrote only about the paint (more often the lack of paint) on the outbuildings and the neatness (more often the untidiness) of the farmyard. Most often, the household member who met with the interviewer was female, and women generally either volunteered no information about the quality of the soil, or they were not asked. (There was one woman who was described as the owner and farmer of her own land, but Anderson's chief comment about the condition of her farm was "lovely view of mountains."[11]) Thus, the interviewers recorded only thirty-four direct answers to questions about soil quality, almost always when men were present for the interviews.[12]

What they heard from those thirty-four men about their land, however, was more than a little surprising: it was strikingly positive. Not a single farmer said his soil was poor. Over half described their land as "good," and another twelve said "fair." They typically used words like "level" or "level in parts." Nine used the specific word "loam" (a very positive term for rich, crumbly soil). A third of the men did acknowledge that their land was stony or hilly, but all those comments were included in overall positive descriptions. Stony soil, it seems, was nothing to fear. One man explained that he had "cleaned up" his stones; another said his land was "very good and productive, though stony." Nor were these farmers daunted by a farm that possessed only "fine patches" of good land, or "fifteen acres of rich level land." A farmer's knowledge of his soil's best uses could go a long way toward dealing with its problems: one field was "very good for corn and beans," another was "good potato land."[13]

To be sure, these farmers were aware that Jamaica's hillside farms posed difficulties they would not have encountered in the broad valleys of Lake Champlain, and they assessed their own land within that context. (One man remarked that his soil was "not especially good," but was at least "more productive than some neighboring farms."[14]) Perhaps there was an element of personal pride or loyalty coloring these statements; it might have been affection for a farm that had been in the family for generations. And given the tenor of the statewide discussion about the future of the hill farms, it would not have been strange if these Jamaica men were feeling defensive about their farms. But their responses do make it clear that they did not think the quality of their soil was in itself the source of their troubles.

The team of geographers, agricultural economists, and foresters who conducted the 1929 thirteen-town study discovered something that lends support to this counterintuitive way of looking at things. In their analysis of 162 sample farms, they were startled to find that the "physical land characteristics" of the farms had little or nothing to do with the crop yields obtained. That is, better quality land in these towns—more level, with deeper and more fertile soil—did not correlate with higher yields. Nor did better land necessarily mean higher income: Peet reported that there was "not a correlation between the farms that had the best land and higher income."[15] He theorized that hill farmers, who often worked off the farm, needed to choose where to concentrate their limited time and effort, and that whatever land received the most attention gave the best results. In that case, the determining factor would be the time available to the farmer, not the quality of the soil. If that was true, then perhaps the "problem of the hill farms" was not simply a matter of stony soil or steep pastures. (This part of the Peet study's results, perhaps predictably, did not make it into the Land Utilization Committee's summary or influence their recommendations.)

Of course, both farmers and experts knew that many hill farms had fields that were so "slopy" they could be worked only with hand tools. There were hayfields bounded by stone walls that could not be enlarged to accommodate the newer haying machines. And there were farms on back roads that could not be reached by the new milk trucks. The experts in the hills sometimes described these problems as if they were fixed, almost immutable features of the land, like bedrock composition or altitude. The 1918 soil survey of Windsor County, for example, reported that there were many areas "not capable of being successfully farmed under present conditions, as the steep slopes and irregular rock outcrops would prevent or greatly hinder the use of modern implements and machinery (See Figure 6)."[16]

The soil scientists recorded that observation, not as historical or social commentary, but under the heading "Soils: Hollis Stony Fine Sandy Loam." But these problems—the slope of the fields, the stone walls, and the dirt roads—are better understood as arising, not from the land itself, but from the demands of modernization. Indeed, they are problems only in that context. Those features of the landscape, both natural and human-made, did little to impede the work of farmers who continued the general mixed farming that had mostly been practiced in Jamaica up to that point.[17] What they did do was make it difficult for hill farmers to turn their farms into the kinds of competitive commercial operations that were now coming to dominate the region's farm economy.

Most of all, modernized farming required one asset completely unrelated to the quality of the soil: capital. Many of the constraints Jamaica farmers faced were more financial than agricultural. No Jamaica farmers said they were held back by the quality of their soil, but several told the interviewers that they would have done better if they had been able to get access to more start-up funds. Even the most fault-finding neighbors acknowledged that this was a significant problem: one woman who excoriated her neighbors for their laziness and lack of ambition had to admit that "it is hard to succeed unless there is money back of them and most of the people haven't the money with which to start."[18]

Most of the farmers who made these comments to the eugenicists were not dreaming of big windfalls. A little would have gone a long way. One man specified that he had good land—mostly loam and good for corn and beans—but he did not have enough money for the commercial fertilizer he needed to supplement his manure.[19] Another, Mr. Nelson, reported that he was saving up to buy another horse. Mr. Nelson's goals were very modest: an elderly man who had worked as a laborer for most of his life, he knew that the small farm he now owned was "stony" and the land was "not especially good," but he hoped a second horse would allow him to expand the amount of land he cultivated.[20] Another man had grander ambitions: Mr. Merrell was young and strong and the farm had "good land—quite level loam." He shared the farm with his brother-in-law, so there was even plenty of labor available. But what they did not have was "enough capital to make [a] living from farming."[21]

One older farmer told Elin Anderson that farming in Jamaica had become more difficult because "the standards of living are so different now." (Attempting to replicate his "good Vermont accent," Anderson took down his words verbatim.) These days, "ye got to go to hear lectures and ye got to have a car—and if a man goes off for a day in his car, it means he ain't workin."[22] This man was onto something, but it was not just the younger generation's love of entertainment that was driving the change he perceived. To operate a new style of farm, a farmer must necessarily "go off for a day in his car." A modern farm must sell milk; older dairy specialties were now increasingly outpriced by midwestern products shipped in refrigerated cars. That crop was labeled "fluid" milk to distinguish it from cream or butter—that is, milk for direct consumption as a beverage. It required speedy transit and more stringent regulation of cleanliness and freshness, and to produce it required off-farm feed and fertilizer and seed. All these things would call for cash—and so, of course, would the car the farmer used to deliver the milk, haul the feed, and travel around to lectures. No wonder Mr. Merrell and his brother-in-law could not scrape together the capital to make it work.

Maintaining Old Ways

Many Jamaica farmers tried to do things differently. Like Nat Price, they continued to operate their farms in ways that had been successful in the past. But that path was not without difficulties, either. The Chipman interview offered Elin Anderson an unusually detailed look at the tensions generated by such an effort. In fact, Anderson seems to have walked into an ongoing argument between the husband and the wife: "a good family with considerable inner friction," she noted. Mr. Chipman—"a big, burly, genial man"—seemed more comfortable with the interview than did his wife, who was described as a "slim young looking fair woman, considerably on the defensive and not very comfortable till the latter part of the interview."[23] (Indeed, it seemed she turned the tables on the interviewer herself, asking "pertinent," but unfortunately unrecorded, questions of her own.)

Mr. Chipman was born in Jamaica; both his father and his great-grandfather had farmed in town. As was common among Jamaicans of that generation, he and his wife moved around for several years working in various jobs before they returned to town in 1914. Although she had initially been the one to insist on settling down in one place, Mrs. Chipman now thought it had been a mistake. She confided to Anderson her fear that in Jamaica, her husband would never commit to what she called "really" farming. Anderson noted: "She now wishes they would either *really farm* or move, but it seems too late now."

At first glance, it is difficult to see just why she thought her husband was not "really" farming. On paper, his farm appears to have been the epitome of a successful, if old-fashioned, operation. He certainly seemed to be doing a lot of work. Over a quarter of his land was in tillage—higher than the town average of 19 percent. ("Tillage" was land that was "tilled"—plowed—and planted to field crops or hay meadow, as opposed to woods or pasture. The percentage of land in tillage was thus a broad indication of how intensively the land was being used.) That spring he produced 180 gallons of maple syrup. He kept a large flock of seven hundred laying hens, "some turkeys," and ten cattle, including seven milking cows, all of which he likely fed with his own hay and corn and other grains grown on the fifteen-acre field he described as "rich level land." Mr. Chipman agreed with his wife that Jamaica was not the ideal place for commercial farming; it was too far from markets. But he was certainly producing for market anyway (hence those 700 hens and 180 gallons of syrup). Why wasn't this "really" farming?

Mixed Crops

Perhaps the wide diversity of crops seemed backward or inefficient to Mrs. Chipman. The farm's mix of maple, dairy, and poultry was not out of line with Jamaica norms. In 1880, it would have been a normal arrangement anywhere in the state. A diversified crop mix was a long-standing "safety first" strategy to protect the households from ups and downs in markets, weather, or other unpredictable setbacks. But by 1930, most of Vermont's farms (58 percent) were classified as "dairy" farms, while just 10 percent were still classified as "general" farms, on which no single crop was the primary income generator. (Only 6.8 percent were classified as "self-sufficing," a new census term to designate the old-fashioned farms on which more than half the value of the production of the farm was consumed by the household itself.[24])

Specialized dairy farming was now the dominant form of farming in Vermont as a whole, although not in the area around Jamaica. Dairy farms were concentrated in the northern and western counties, with the highest percentage found in Franklin County in the northwest corner, and the lowest in the southeast corner—Windham County, where Jamaica was located (see Table 8). The old-style "general" farms, in contrast, were most common in the two southernmost counties, as were those labeled "self-sufficing." A good many of those "self-sufficing" and "general" farms were no doubt to be found in Jamaica and its neighboring hill towns.

In the northwestern counties, farmers were increasingly heavily invested in a kind of commercial dairy farming that would ultimately lead them far away from the old mixed farming patterns of the past. In 1930, to be sure, they were not yet all the way there: most of these farms were still more diversified than they would become later. But more and more of their energy was focused on their dairy herds. Farmers used more machinery in both field and barn, bought more off-farm feeds and fertilizers, and borrowed more money from the bank. Herds were becoming larger. Black-and-white Holstein cattle, bred for maximum milk production, were beginning to replace the fawn-colored, large-eyed Jerseys and Guernseys whose richer milk had been better suited to the production of butter and cream, along with the red Devon multi-use cattle that had been the first cattle raised in the region. (One midcentury commentator jokingly accused some Vermont farmers of "clinging to the Guernsey and Jerseys of their fathers."[25])

Perhaps, then, Mrs. Chipman hoped her husband would become a dairy specialist of that kind. Theoretically, that would not be impossible: four out of five farms in Jamaica did produce some dairy products, Mr. Chipman's farm among them.[26] But most of Jamaica's farms showed few signs of moving toward modern commercial dairy production. A comparison with Cornwall and Waitsfield, the other two towns studied by the eugenicists, makes this clear.

Table 8. Dairy Farms, General Farms, and Self-Sufficing Farms by Percentage*

	DAIRY**	GENERAL	SELF-SUFFICING
RHODE ISLAND	35%	11%	10%
CONNECTICUT	32	10	10
MASSACHUSETTS	30	10	7
NEW HAMPSHIRE	29	15	12
MAINE	17	19	10
VERMONT	58	11	6
FRANKLIN COUNTY	75	5	3
ORLEANS	72	8	2
CHITTENDEN	67	5	6
GRAND ISLE	65	8	7
ADDISON	60	9	5
CALEDONIA	61	7	4
LAMOILLE	60	13	2
RUTLAND	58	7	8
ESSEX	57	14	8
WASHINGTON	56	12	4
ORANGE	52	13	5
WINDSOR	47	4	8
BENNINGTON	40	17	13
WINDHAM	34	20	11

*These do not add up to 100 percent because there are other categories of farms not discussed here.

**The farming income categories include households who reported mixed income from both farm and other sources.

For one thing, Jamaica herds were markedly smaller than those in the other two towns (see Table 9). In Waitsfield, the average herd size in 1930 was twenty-two cattle, and in Cornwall it was eighteen—both above the state average for census-designated dairy farms. In Jamaica, the average was just a little over six—the statewide average for "general" farms. In both Waitsfield and Cornwall, larger herds dominated: nearly two-thirds of those who owned cattle had ten or more. In Jamaica, the situation was reversed: four out of five farms with cattle held fewer than ten. Mr. Chipman held ten cattle: near the top rank of farms in Jamaica, but in the lower middle rank in Waitsfield or Cornwall. At the lowest end of the spectrum, over a third of cattle owners in Jamaica owned just one or two cows. At the upper end, over a third of Waitsfield farmers held thirty or more. These were obviously not the same kinds of farms.

Table 9. 1930 Cattle Holdings in the Three Towns of the Migration Study

#CATTLE/ FARM	WAITS- FIELD	WAITS- FIELD %	CORN- WALL	CORN- WALL %	JAMAICA	JAMAICA %
1	10	12%	11	12%	17	22%
2–9	19	23%	23	25%	45	59%
10–19	14		29		10	
20–29	9		9		1	
30–39	9		11		3	
40–49	8		4		0	
50 and up	12		6		0	
10 and up	52	64%	59	63%	14	19%
TOTAL FARMS	81		93		76	

In Jamaica, moreover, half the farms selling dairy products (including Mr. Chipman's) were still producing butter or cream—old-fashioned products for declining markets—rather than converting to the newer fluid milk market.[27] And some Jamaicans who raised dairy cattle were hesitating to convert to the sale of milk for a reason that harkens back to the earliest farming practices in town: they were cattle traders more than dairy producers. Farmers whose primary interest was in raising stock preferred to leave the young animals with their mothers for feeding rather than taking all (or perhaps any) of the milk for sale. A pair of brothers explained to Anna Rome, for example, that they sold little or no milk because they used almost all of it to feed their calves.[28] Another farmer reported that he had decided to "begin shipping milk rather than cream," but that he hated to do it, "for then he won't have any for his calves."[29] (While he was shipping cream, in other words, he had been feeding his calves the skimmed milk.) Another couple explained why that choice made sense: "They do not consider this a loss of money," Rome explained, "as they feel that they can sell the cattle for much more."[30]

Even on Jamaica's largest and most profitable farms, those older local preferences were still evident. Mr. Graham, for example, was reputed to be the wealthiest man in town. "Mrs. G says they make $1000 or $2000 a year but was not at all certain," Marjorie Choate reported, adding that they made enough money "so that they have to pay an income tax,"—a standard only one out of four hundred American farmers had reached by the first assessment in 1916.[31] A more vivid testimony was the fact that Mr. Graham was the only person in town with a milking machine—an important step in modernizing dairy production.

It was a big operation by Jamaica standards: Mrs. Graham estimated that the farm currently held somewhere between thirty and thirty-five cattle. But only seven of those cattle were milk cows. Mr. Graham did not specialize in dairy: like Jamaica farmers back to the first settlement of the town, he traded in cattle. In summer, his wife reported, he sometimes kept as many as a hundred head of cattle at a time on his farm. His two brothers also farmed in Jamaica and employed a similar strategy—"speculating in cattle." They had learned the trade from their father, who had been a cattle dealer, too.[32]

To be sure, dairy products were the biggest cash generators for most Jamaica farms both large and small. But that fact did not lead most Jamaica farmers toward specialization in dairy production. On the contrary, most Jamaica farmers, big producers as well as small, were still operating "general" mixed farms like Mr. Chipman's, producing a variety of goods for market. Mr. Livingston, for example, was described by Anna Rome as "one of the most outstanding farmers in town."[33] He and his wife made a good living in 1930, bringing in over $2400, but milk provided only a third of that income. The rest came from a wide assortment of other crops: maple syrup, cattle, eggs and strawberries. Low-income producers were often equally diversified, often by necessity. Mr. Williams, for example, was a sixty-nine-year-old widower living alone. He patched together a living tapping maple trees, "working out" for other farmers, picking ferns (a popular specialty for Jamaicans, responding to a demand for fresh ferns from urban florists), and selling berries from his large garden. He told Marjorie Choate that he kept no yearly accounts at all but thought his total cash income was about $200.[34] His hands were "crippled with rheumatism so that he cannot milk," or he would have kept a cow like everybody else.

Accounting

All this diversification made for complicated—or more often nonexistent—recordkeeping. On the printed form used by the interviewers, there was space to enter both "gross" and "net" cash income, and the interviewers did record something in at least one of those spaces for most residents, but the diversity of answers makes it clear that Jamaicans were not usually thinking in precisely the terms the interviewers were looking for. Mr. Chipman reported to Elin Anderson, for example, that he had made $350 from maple syrup and $450 from cattle (including both cream and beef). His responses suggest that he understood what "net" and "gross" meant: he subtracted an off-farm grain bill of $250 from his total for the cows.

Even farmers who gave net figures for their incomes counted only a few expenses, typically subtracting the costs of materials for which they had paid cash—most often grain for dairy cows. They did not include their own labor, upkeep to the farm, or other types of expenses an accountant or farm economist might have suggested to them. Mr. Chipman also neglected to report any income he might have derived from his seven hundred laying hens, nor did he count the turkeys in his final figures, perhaps because he had not sold them yet in early November. (It is possible that he did not count the egg money because it accrued to his wife, as egg money sometimes did, but the eggs produced by seven hundred hens is a lot to leave out of one's accounts.) And when Anderson, adding it all up, asked him to confirm that he must have cleared about $1,000 that year, he agreed: He knew that his family had *spent* about $1000 that year, so he supposed they must have made that much, too.[35]

That may sound at first like a little joke, but the remark likely had a meaning that is not immediately apparent. The thirteen-town team had encountered something similar in their study of hill farms the previous summer. The team tried to calculate the yearly profits of the 162 households they had chosen for close examination. After a long and meticulous process, they concluded that these farmers basically made no profits. They did not carry much debt, but neither did they accumulate much in savings.[36] The farm itself served as bank account, financial portfolio, job, and pension rolled into one, but it was not something they could cash in on or alter without endangering the household's security. (That was part of the reason why very few farm owners in Jamaica or any other farm towns studied were the ones to sell out and leave town: there was not much chance of recouping all that was invested in the farm.)

Instead, as Peet explained, their "living expenses were carefully adjusted to equal that which was available for spending."[37] That, too, may sound like an obvious remark—they spent only what they had—but to the experts, this way of operating did not mean simply that the households were living within their means. It meant they were not "getting anywhere," or as Peet wrote, they were just "beating time."[38] Worse, it implied that they were not *trying* to "get anywhere." Perhaps it even meant that these families were not thinking about their lives as a forward trajectory at all, but rather as a seasonal cycle with nowhere to "get." Maybe that was what lay beneath Mrs. Chipman's disapproval.

Mr. Chipman's income figures were actually a good deal more precise than many in Jamaica. It seems likely that many of the figures the residents gave to the investigators were more like general estimates than precise bookkeeping. One man told Anna Rome that he had made $500 that year, but he added that he really had "no idea what is taken in."[39] Indeed, a suspiciously large number of residents seemed to think that their net income was $500. Nine households in all gave the figure of $500 as their net income, while no other number was reported by more than two households. Perhaps $500 struck a lot of people as a respectable income.

The imprecision of Jamaica farm accounting seems not to have been because residents were incapable of handling the arithmetic.[40] Nearly all Jamaica residents had attended district schools.[41] Nor does it necessarily mean that Jamaica farm households handled little cash or that they did not know the market value of their products. Like their parents and grandparents before them, Jamaica farmers seem to have been keenly aware of rising and falling prices in local and regional markets. They could be innovative in the pursuit of more income, experimenting with new crops and methods when their situations allowed. (Several examples of those experiments were recounted in the eugenicists' field notes.[42]) And they could be meticulous in the accounting of the financial transactions that mattered. But they were not closely focused on the net profit and loss figures of the operation taken as a business. Even one of the prosperous cattle trading Graham brothers told Marjorie Choate that he "regretted that he did not keep accounts so that he did not know what his income was."[43]

It may be that the question of "net" annual cash income was not a very useful one for farmers who mixed subsistence and market crops, wages and sales, as many of these households did. In her notes about Mr. Williams, the man with the arthritis, for example, Choate reported that his yearly income was "probably about $200—keeps no accounts." After that she recorded the rest of what he had told her: figured in wages, he earned $2.50 per day for working off the farm; measured by the quart, he received $30 per hundred gallons of maple syrup; at a rate of $7 per thousand feet, he sold ash logs from his property. It would become even more complicated if he attempted to keep track of the pints of strawberries he sold to neighbors; the bushels of potatoes he stored in his cellar; or the number of ferns he sold to the wholesaler. This man had no trouble counting or keeping track of numbers, but while he kept those figures in his head, it would be difficult to add them up in a way that gave him any useful information.

No Jamaica resident broached this subject, but perhaps there may also have been a reason rooted less in practical usefulness and more in deeply rooted cultural practices. An advisory essay for census takers in the introduction to the 1900 agricultural census described at length the "utter indisposition" of the average farmer to figuring profit and loss on goods produced for and consumed by the household. It was "altogether alien and repugnant" to farmers to "give a value to the garden truck that is carried into the house, the fuel picked out of his woods, the fruit that his children eat, the corn that is sent to the mill for home use, or even the pig that is killed at Christmas." And of course, it might be difficult to separate the children's apples from the apples for market, or one's own Christmas pig from the ones sold to one's neighbors. The essay attributed that deep repugnance to figuring profit and loss to the farmer's belief that these goods were an essential part of a kind of "contract" with the farm.[44] It would be a denial of the bond between farm household and farm, or of the farm's ability to support its people.

For whatever combination of reasons, Jamaica farm households were conspicuously different from the other two farming communities studied by the eugenicists (see Table 10). They were willing and able to add up what cash they had received for which products, but beyond that many were unwilling or unable to go. A little over a third of farming households in Jamaica gave the investigators a figure that they reported as "net" income, while almost two thirds used only gross figures.[45] In striking contrast, just five farm households in Waitsfield described their income using a gross figure, while over ten times that many used net figures. Similarly, in Cornwall, four times as many households used net figures as gross. Perhaps no other comparison of the three towns illuminates with such clarity the different path most Jamaica farmers were embarked on.

Table 10. Farmers Reporting Income in Gross and Net Figures

	FARMS USING NET FIGURES	FARMS USING ONLY GROSS FIGURES	AVERAGE NET INCOME	AVERAGE GROSS INCOME
JAMAICA	22	43	$573	$770
CORNWALL	46	11	$1,042	$1,523
WAITSFIELD	54	5	$1,602	$3,832

"Working Out"—Off the Farm

Diversified crops and rudimentary recordkeeping were characteristic of the kind of traditional mixed farming that most northern farmers had practiced for generations. In 1930, they might well have seemed old-fashioned or out of step with modern times. But it was a different feature of Mr. Chipman's business practices that drew his wife's most open disapproval: his decision to combine his farm work with working off the farm—referred to in the field notes as "working out." She thought that her husband "could make good if he really farmed," Anderson wrote, "but she says he's more interested in blacksmithing etc." Chipman's grandfather had been a blacksmith, but as Anderson pointed out in her notes, it was not a craft that was in much demand by 1930: "certainly there is no thought of his doing it now." In this case, too, Mr. Chipman was in good company: just as most Jamaica farmers still relied on a variety of cash crops, combining farm work with other kinds of labor was pervasive in town. It was standard practice in the thirteen hill towns studied by the Peet team as well.[46]

In the past, combining farming with non-farm jobs had been common, too—just another way to diversify the income streams of the household. Farmers operated all kinds of artisanal sidelines, with places of business often located on their own property, like the Prices' wheelwright shop. The 1884 Child's *Gazetteer* lists several Jamaica farmers who were carpenters and cabinetmakers; as well as two coopers, a tanner, a stone mason, a watch repairer, and a shoemaker. By the late nineteenth century, industrialization had begun to depress demand for some of these trades, and by 1930 almost all were gone.

A few farmers were still able to develop profitable sidelines. Mr. Chipman did blacksmith work when he could get it. Mr. Baker sold fire insurance.[47] Mr. Williams, whose arthritis made it difficult for him to milk cows, had acquired the skills over the course of his life to work as a barber, make and repair harnesses, and even to do "some undertaking" should that be required.[48] One couple ran perhaps the most diversified farm in town, including both several market crops and several non-farm jobs. The foundation of their farm was a successful "boutique" poultry operation, sending eggs by express train to Boston. On the side, they sold butter, canned vegetables and fruits, and hosted summer boarders. In winter, Mr. Halvorson also worked as a sign painter, but his most lucrative sideline was making violins ($75 to $400 for "two weeks' steady work"). Just as they did with their eggs, the Halvorsons sent the violins to Boston.[49] All told, they made a combined income of $1,830 that year. Mr. Halvorson was a high school graduate, but he reported his income as a gross figure. It would take a skilled bookkeeper to keep track of income and outgo from an array of businesses like these.

For those without such an array of specialized skills, there were still other options in town. Logging had been a common winter occupation in the 1880s, and those jobs were still around in 1930.[50] There was still some work available in the four or five sawmills left in town, although less than there had been a generation earlier (and much less at the onset of the Great Depression in the fall of 1930). The newest sideline was "ferning"—picking ferns for florists to use in flower arrangements in the city. A 1923 *Brattleboro Reformer* headline described the fern "boom" and the easy money it was bringing to the region: "Thousands of Dollars Being Spent in Vicinity of Jamaica."[51] The newspaper story reported that the right kind of ferns (Boston and lace ferns) were found "in the deep woods of the mountains above the headwaters of the little West River," in Jamaica and a few surrounding towns.

Like other crops, ferns had good years and bad years, but good pickers could earn high wages during the weeks of summer when the ferns were harvested: The *Reformer* article noted that Mr. Williams, described as a "bearded Jamaica farmer," had delivered "3000 ferns in his wife's clothes basket." The ferns were worth two-and-a-half cents per bunch of twenty-five, and thus would yield him about three dollars, better than his wages for a day's labor on someone else's farm.[52] Whole families packed up picnics and went up to the hills to camp in their favored spots. Several men in town ran lucrative wholesale operations. One told Marjorie Choate how he had developed a "fern jobbing business of tremendous size," employing forty people in town in the season. At that size, it is difficult to say whether the fern operation or his farm was the actual sideline.[53]

"Working Out on the Roads ... Etc."

"Ferning" was the newest part-time job for many people in Jamaica, but the oldest and most reliable year-round off-farm work for Jamaica farmers was what might be referred to now as infrastructure repair. That included both the never-ending work on the town's roads (snow packing, brush clearing, ice breaking, gravel spreading, re-grading, clearing drainage culverts ...) and larger projects such as repairing the railroad or the fifteen bridges in town. So many Jamaica famers worked on those projects that by the time Anderson reached her late November interviews she had begun to note the pattern in shorthand: "Works out on road etc. besides farming ... etc. [her ellipses]."[54] In fact, she disliked the practice so much that in her final report she commented acerbically that it was "fortunate" that "the flood of 1927 did considerable damage in the town." That was because all that destruction had made necessary "the repairing of roads, the building of bridges and the rebuilding of the West River railroad," which "have provided employment for many people ever since."[55]

Anderson's "fortunate" remark was insensitive, of course, given the catastrophic damage caused by the flood of 1927 not only to local roads and bridges, but to farmland, shops, schools, roads, and houses across the state (to say nothing of the eighty-four people killed).[56] But perhaps her hardheartedness should be excused. From the standpoint of all the many investigators in the hills, labor on the roads was indeed one of the worst ways Jamaica farmers could have chosen to deal with their narrowing economic options. Farmers had always worked on the local roads, but now agricultural experts viewed their reliance on off-farm jobs as a wrong turn in the road to modernization, a regression to more primitive ways of organizing rural life. Budget-conscious state leaders worried that the dwindling population of the hill towns, combined with the need to keep the roads passable even in remote parts of town, was resulting in unsustainably high taxes spread among fewer and fewer taxpayers. The state was already footing the bill for some of the hill towns' road repairs and school expenses, and they feared things would get worse. Road work seemed undesirable from a eugenical standpoint, too: if the town paid its residents to maintain back roads that served increasingly few people, it encouraged both the workers and the residents on those roads to stay in town, rather than forcing them to go somewhere else.

The eugenics team also suspected that Jamaicans' combination of wage and farm work was somehow related to their passivity and lack of ambition. Perhaps this is how Mrs. Chipman viewed her husband's mix of jobs, too. A handful of Jamaica residents saw "working out" as a sign of their neighbors' laziness, or even that worst of all judgments, "shiftlessness."[57] Elin Anderson spoke at length with one Jamaica resident, Mr. Strong, who offered a similar but somewhat more nuanced interpretation. Farmers who "worked out," he explained, were not really lazy. Rather, they were the victims of a kind of addictive cycle they were unable to escape: "When it comes spring and time to put in the crop, they need a little ready money to buy feed, etc. So they go out and work on the road for a few days." But while they are working on the road, "they forget all about their farms." (Mr. Strong was the man who described his neighbors as lacking "planfulness"—the type of people who might "forget all about" a farm.) Then, "before they know it," it is too late to plant, and they find themselves forced to continue to work on the road to bring in an income.[58]

Mr. Strong's explanation was no doubt well meaning, but much more straightforward explanations are readily available in the field notes themselves—explanations that do not require belief in the existence of a group of farmers so "planless" that they forget all about their farms or lose track of the seasons. Mr. Chipman, for example, made his situation quite clear to the interviewers: he simply found it impossible to make a living for his family of nine from his farm alone. The thirteen-town study found that a herd of ten cows like the one on Mr. Chipman's farm was the bare minimum "necessary to maintain more than a 'mere existence' in life." Even then, there would be a need for "supplementary sources of income."[59]

Farm products now competed on broad regional markets, not local ones, and as the thirteen-town study pointed out, Jamaica was "far away from markets." The Halvorsons worked around that disadvantage by using an expensive express train that would get their eggs to a high-end urban retail market. Mr. Chipman took a different path to his "supplementary sources of income," cobbling together a decent living despite the market disadvantages he faced. In 1930, he did it by tapping maple trees ($350), raising calves and milking cows ($450), and also by that blacksmith work his wife so disliked ($450).

The patchwork of off-farm and on-farm work Jamaicans developed in these years was perhaps not ideal, but it was workable. Despite the Chipmans' disputes about methods, they lived on the farm until Mr. Chipman's death in 1944. (In fact, all seven of their children were still living in Jamaica or in adjoining towns at the time Mrs. Chipman died some twenty years later.[60]) It was a system that made possible a continuation of the household's traditional "safety first" diversification and avoidance of dependence on a single source of income. That system allowed the household to take care first of its own basic needs regardless of the vagaries of the labor, housing, or food markets. And it permitted them to use the land they had bought or inherited in mostly familiar ways, though often less intensively than before.

More Work than Farm

In one sense, though, Mr. Strong was right. It was probably not that anybody "forgot all about" their farm—but that the need for off-farm work often did push farming to the side. Indeed, the whole situation was almost as circular as Mr. Strong made it seem: the thirteen-town study suggested that good land did not necessarily translate into good yields or high profits in the hills because the farmer could not spare the time to farm intensively; the farmer's time was short because he needed to work off-farm; he needed to work off-farm because his farm could not make a living in the modern economy. No matter how "planful" a farmer might be, there were only so many hours in the day. For that matter, much of the wage work in Jamaica was available only in summer, when farms, too, demanded maximum labor. Mr. Chipman had two important advantages: his seven children, ranging in age from 10 to 23 in 1930, no doubt helped out on the farm; and his skilled blacksmith work earned more than twice the going daily wage of work on the road. Many farmers could not juggle off-farm and farm work so effectively and were forced to give less attention to their farms.

Anticipating the difficulty with that strategy, Mr. Merrell laid out his whole plan for Elin Anderson. Merrell and his brother-in-law were the ones who told Anderson that they did not have access to enough capital to increase their production to a level where farming would provide a good living. Recognizing that fact, the two men decided on a different path: both took jobs, one on the bridge and one on the railroad. They kept the farm—that part of the plan was essential. But they gave up on producing for market. They would work for wages to get the cash they needed, and the farm would continue to provide the food, shelter, and heat that would lessen their reliance on inadequate or unreliable wages. (In fact, both jobs were scheduled to end just around the time they spoke with the eugenicists.) In the interviews, that set of choices was often labeled "home use."

Home Use

It was a deceptively simple phrase. "Home use" might refer to any one of a range of farming possibilities. At one extreme, it might mean that the farm provided only for the needs of the household—that it was a subsistence or "self-sufficing" farm, in the new census terminology, producing no crops for market. One Jamaica resident complained that it never even occurred to her neighbors to try to "grow more vegetables than for their own use nor even more cows than are needed to supply their own milk."[61] But Mrs. East was prone to exaggeration: not many farmers in Jamaica in 1930 operated strictly "self-sufficing" farms (nor had many farms operated that way a generation or even two generations earlier).

"Home use," then, might also mean that the farm was *mostly* dedicated to providing for its own household but that it regularly marketed its "surplus"—sold the products it had produced more than enough of. A generation earlier, that would not have required a special label at all. It would simply have been called "farming." It was conventional wisdom, as an 1884 report put it, that a hill farmer should "aim to supply his own necessities, as far as practicable, from the farm, and for the surplus raise that which will bring most for the least expense and not impoverish the farm."[62] In 1930, there were still many farmers in Jamaica who operated that way (see Table 11).

Anna Rome described the Bauers, for example, as cultivating "only enough for home use," yet she also recorded that they sold an estimated $300 worth of maple products and $200 "from the cows" that year. In other words, the Bauers farmed mostly to supply their own needs, while also marketing some of their cream and producing maple syrup as a cash crop (and presumably also keeping some for themselves). The $500 would have provided the cash they needed, but it was only a part of what they had to live on.[63] Even in the more "progressive" towns studied by the eugenicists, at least a few farm households were producing mostly for "home use" in that sense of the term. In ultra-modern Waitsfield, there were just three households labeled by the interviewers as "home use," but in Cornwall, a more traditional farming community, there were fifteen (9 percent of the households in town—a little higher percentage than in Jamaica). In the broadest sense of the term, moreover, growing crops for "home use" simply meant that in addition to whatever market crops they specialized in, and no matter how large a commercial operation they ran, farmers still produced as much as they could for their own consumption. That practice was not (yet) old-fashioned or outdated anywhere at all: it was a widely practiced strategy even in the most commercially oriented farm households in Vermont.

In 1929, a team of home economists at the Vermont Agricultural Experiment Station launched an intensive study of its own—a year-long analysis of the food budgets of forty-three farm households from eight different counties in the state, mostly from the more rapidly modernizing dairy regions.[64] These farms were certainly not "self-sufficing," but they did supply on average 95 percent of their household's milk, cream, and cheese; 88 percent of the eggs; nearly half of the meat; and 71 percent of the vegetables and fruits. In terms of dollars, on average these households provided 54 percent of their food from the farm.[65] And they had good reasons for continuing to do things that way, as the home economics team discovered. The families that produced most of their own food consistently reported the most nutritionally complete diets, while the poorest and least well-nourished were those who provided the least of their own food.[66]

Table 11. Farming for Home Use

	CORNWALL	WAITSFIELD	JAMAICA
All households with information	157	191	159
All households growing food as percentage of all households	144 (92%)	128 (67%)	136 (85%)
Households growing for "home use only" or "just gardens"	22 (15%)	12 (9%)	32 (24%)
Farms >20 acres cultivating <5 acres as percentage of food growers	7 (5%)	9 (7%)	30 (22%)

Food self-sufficiency was evidently still a good economic strategy for most rural households in 1930. Even prosperous modern farmers relied on food from their gardens and barnyards to get them through the winter. Onions and squash were stored in the attic or an unheated room. Beets, carrots, and cabbage were stored in root cellars along with salted meat and the indispensable potatoes. (The home economists estimated that Vermonters required somewhere between a hundred and two hundred pounds of potatoes per person per year.[67]) The home economists also reported that farm households at all levels of production "put up" large quantities of canned fruits, vegetables, pickles, and jams for winter consumption (see Figure 7). In Jamaica, one woman explained to Anna Rome how that practice might pay off for a household that relied on farming to supplement unreliable or poorly paid wage work. To be sure, she and her husband were "worried about winter since there is so little outlook for work." But she was confident the family would get through: "Mrs. R says they have plenty of canned foods and potatoes to carry them through the winter."[68]

Just Gardens

In fact, without stretching a point too far, one might place most Vermonters in the category of those who produced food for "home use." The home economics team found that even among the non-farming population in the towns and villages, almost three quarters of households raised at least some food for themselves.[69] In 1930, 67 percent of the households of relatively urban Waitsfield raised some food, if only a cow, a garden, or a flock of chickens. On the broad open acres of Cornwall, 92 percent of households produced some of their own food. In mountainous Jamaica, in far more challenging circumstances than those in Cornwall, 85 percent of households produced some food for themselves. Only

FIGURE 7. "Farm Security Administration client with canned goods." Farming for "home use" was a mainstay of hill farm life, including preparation and storage of homegrown food for winter. While the eugenicists took little note of subsistence practices like these, by the time this woman was photographed a few years later, New Deal agricultural specialists were enthusiastically encouraging farm households to continue or reinstate such old-fashioned practices, providing for their own "home use" by canning and cellaring food for the winter. Source: Farm, Bradford, Vermont, Orange County. Photographer Leo Russell, 1939. Library of Congress, Prints & Photographs Division, FSA/OWI Collection, LC-USF34- 034323-D [P&P] LOT 1235.

23 households out of the 158 in Jamaica reported that they engaged in no food production at all.

The interviewers used the phrase "just garden" to distinguish small producers from those who were operating full-size farms—but that designation, like "home use," was not always what it seemed to be. The gardens the interviewers described as "just gardens" were never less than half an acre and usually much larger: in Waitsfield, a seven-acre garden plus one acre for the chickens; in Cornwall "merely a two-acre garden;" in Jamaica, a four-acre garden that kept the ten people of the household "well supplied with garden stuff and potatoes."[70] (An acre is a little less than one American football field, or about fifteen tennis courts.) A garden of this size was really something like a subsistence smallholding: it would produce much more than fresh vegetables for the summer months, including nearly everything needed to feed the family for the year, and sometimes some to sell to neighbors. Many such gardens also grew at least part of the feed for the cow, chickens, or other stock. On one four-acre plot in Cornwall, for example, the resident raised three acres of corn and potatoes, along with a "small" one-acre garden and a flock of chickens. Within this context of nearly ubiquitous production of food on a small scale, there was nothing anomalous about the "home use" strategy of Jamaica households.

Jamaica *Was* Different

Still, the interviewers did have a point. There really was something different going on in Jamaica. The team frequently used the term "just garden" to refer specifically to those farmers, like Mr. Merrell, who were reducing their production for market in order to work for wages. There were many such farmers in Jamaica. In relatively flat, fertile Cornwall, most people growing food were operating full-sized farms, rather than smaller "just garden" plots. In Waitsfield, many residents maintained gardens on small plots of land, as one might expect to be available in a bustling town center. What was remarkable about Jamaica was the large and growing number of residents who owned full-sized farms but raised "just gardens." The Towers, for example, owned eighty acres, but cultivated only a half-acre garden plus hay and grazing land for two cows—as Mrs. Tower expressed it to Elin Anderson in her interview, "just enough for ourselves." Mr. Tower worked full-time at a lumber mill, or at least he did that when the mill was operating. At those times (not very often that year), he had little time for more than a vegetable garden, potato patch, and a few cows and chickens.[71]

At least thirty households in Jamaica fell into this category: farm owners like the Towers, who were cultivating subsistence plots on full-sized farms. Even at that conservative estimate, the number of households farming in this reduced way was far higher than in either of the other two towns, where such practices were quite rare. Only seven households in Cornwall and nine in Waitsfield were using properties of over twenty acres primarily for gardening and a few animals.

And there were also many more households in Jamaica who were making a different kind of use of their farmland.[72] Some households were still marketing the products of their land but shifting away from the traditional cash crops of potatoes, corn, and dairy, toward a reliance on cash crops from the woods. The thirteen-town study found that nearly half of the 162 farms it examined sold maple syrup, and a quarter sold timber from their land. Including the value of the wood harvested on nearly all farms for household heating, the combined average income of products of the woods was $381 per farm—a substantial contribution to most farm incomes in the hills.[73] Some Jamaicans went even farther, relying almost entirely on ferns, hunting, or trapping for their living. One pair of brothers reported to Anna Rome, for example, that they did a little farming in summer, and trapped in winter, but ferning was their chief source of income. They had cleared a substantial $400 each from a few weeks of ferning that year.[74]

Missing the Point

Some Jamaicans gave the interviewers detailed explanations of all these strategies: some "just gardening" and working for wages; others patching together a mix of crops and jobs; some harvesting the woods; still others relying solely on "self-sufficing" farms. But the team often missed or misunderstood those explanations. In one curious example, Marjorie Choate interviewed a man just a few weeks after his family had returned to Jamaica. Years earlier, Mr. Andrews had operated this same farm as a dairy, selling cream, but the family had moved to Massachusetts when he found work there. Now in his fifties, he had lost that job, and they had returned to the farm. This time, he told Choate, he planned not to "run the farm for profit" anymore, instead growing most of what the family needed and trading their surplus with the local store for the rest.

Mr. Andrews explained to Choate that the creamery nearest to his farm was no longer in operation. Many hill towns were experiencing that loss as Vermont's farming infrastructure shifted from supporting butter and cream production to milk. Milk processing plants were centralizing and closing the creameries in smaller towns—one of the many ways hill farmers were increasingly at an economic disadvantage. Losing the creamery could be a serious blow to the remaining farmers in the area. Mr. Andrews calculated that transportation costs and travel time would make it difficult to make a profit using the closest creamery remaining, in Brattleboro, twenty miles away. Yet he valued the basic food and shelter the farm would provide his household as he aged, and as hard times loomed ahead. Choate summarized his explanation: "He thinks the man who stays on the farm is much better off than the man who goes to the city."

For all that, however, Choate seemed unwilling to accept his explanation. Instead, she fell back on an explanatory framework familiar to eugenicists, although not frequently employed by this team of investigators. In three separate places on the interview sheets, she wrote that the Andrews family were "Roman Catholics [French]," "French Canadian," and "French Canadians and Catholics." Finally, under "Summary," she noted their ethnicity for a fourth time, this time using the word the team had hit on to characterize Jamaica's unique social problem. They were "A French Canadian family who seem content just to be back on the farm."[75]

Mr. Andrews would probably have been startled by this interpretation of his actions. He had given Choate a full account of his situation, and she had duly written it down. Her response seems oddly out of place. Perhaps it was an expression of her discomfort with the types of farming adaptations the team encountered in Jamaica. As with the phrase "perfectly contented," the labels "home use" and "just garden" sound innocuous at first, but in context, they conveyed deep disapproval. It is easy to believe that the interviewers found the situation in Jamaica disheartening. One way or another, Jamaica's farmers were now producing *less* for market than their parents and grandparents had done. The fields their grandparents had cleared with such backbreaking toil were growing back to woods. Although they had sound reasons for operating this way, it must have seemed to the interviewers like a regression to a more primitive way of life, even as the rest of the state was loudly declaring its commitment to progress.

But the eugenicists misunderstood something important about the varied strategies these farmers were employing. They were not returns to the past; rather, they were adaptations in their own right to modern times. The decision to farm less was not a sign of giving up. It was in fact almost the opposite—a decision to keep farming. For many Jamaicans, that was the only way to remain on the farm.

How to Count a Farm

The eugenics investigators agreed with Mrs. Chipman: they did not think those Jamaicans were "really" farming. Out of 158 households, they classified only 42 as farms—just 26 percent of the households in town.[76] Those were households that conformed to at least some of their notions of how Jamaica farmers ought to be evolving with the times. But the eugenicists were not the only ones counting farms that year. Just a few months earlier, the federal census enumerator had recorded not 42, but 102 farms in Jamaica—amounting to nearly two-thirds of the households in town. Such an enormous discrepancy suggests that fundamentally different definitions were in play.

The 1930 census form allowed the enumerator to indicate that a place was a farm in three different ways: checking a box to indicate that the place was a farm; designating the head of household as a "farmer" in the occupation box; and—if the place met the minimum requirements—assigning it a number that meant that it would be included in the agricultural census. The requirements were simple: a farm was any productive land over three acres. (If the farm was under three acres, it must produce more than $250 worth of goods, but the definition did not specify how productive the farm had to be if it included over three acres.) A census enumerator could mix and match the categories as seemed appropriate, but the place was not officially counted as a farm on the agricultural census unless it was given a schedule number.

The eugenicists did not establish a specific definition of "farm," but they had a clear idea about what it was *not*. They did not count a place as a farm if the head of the household worked at some other occupation. Mr. Baker, the 1930 census enumerator, took the opposite approach, assigning numbers for the agricultural census to thirty-four places where the heads of household reported non-farming occupations.[77] But even that does not fully account for the discrepancy: there was still a difference of seventy-three farms between the two counts. The question was not only what types of work the head of household performed in addition to farm work; it was also what the household was doing with the land. Census-takers were instructed to count all the land that the farmer considered part of the farm, including "considerable areas of land not actually under cultivation and some not even used for pasture."[78] The eugenicists tried to exclude places they judged to be under-used as farms. Mr. Baker made no such distinctions. In his account of the town, all kinds of people owned farms: landscape painters; village laborers; even the owners of the mills they worked in. And with a single exception (for the pair of brothers who made a living ferning), every residence he recorded as a farm was also awarded a number officially including it in the agricultural census.[79]

Mr. Baker himself was a member of a distinguished local family, the son of a former state senator and the nephew of a state Supreme Court judge. Over the course of his life, he would be a representative to the Vermont state house and serve as a town selectman for forty years. But he was also deeply embedded in Jamaica farm life. He was a farmer who had inherited the home farm on his father's death. As a sideline, he sold fire insurance. When Elin Anderson interviewed him for the Migration Study, she reported that he had "always liked it" in Jamaica "quite well" and did "not want to go away." Indeed, he "probably would not be happy anywhere else than here."[80]

Perhaps Mr. Baker's remarkable census numbers reflect something of his community's understanding of what farming meant. After all, the census training manual instructed the enumerator to count a place as a farm if it was "locally regarded as a farm."[81] That seems straightforward enough, but Mr. Baker seems to have been unusually conscientious about adhering to community standards. The man who had conducted the 1920 census was similarly well connected in town, yet he had counted only seventy-eight farms back then, a number much more consistent with the farm counts in 1940 (sixty-three) and 1950 (sixty-seven).[82] Perhaps the best explanation for why Mr. Baker was so generous with his farm designations is that he simply took his neighbors at their word: that is, he recorded 102 scheduled farms because 102 residents told him they owned farms. In a small way, Mr. Baker seems to have been expressing his solidarity with his neighbors by recognizing their status as "real" farmers—even if they worked off the farm at other occupations, even if they raised food primarily for their own household's consumption, and even if they worked more in the woods than in the fields.

It was all too much of a stretch for the eugenicists. Anderson recorded her frustration in an interview with a man who told her his father had "farmed it" in Jamaica back "when this was quite a busy place." Perhaps thinking of all those 102 Jamaicans who still claimed to be farming, she seemed to throw her hands into the air: "just what 'farming it' means in this town is more than I yet know."[83]

CHAPTER 5

Standards of Living

All the experts agreed: the people of the hills were poor. To use the term that had recently come into favor, hill farmers had a notoriously low "standard of living." One goal of all the surveys was to determine just how low that standard of living was. And in some ways, they confirmed everyone's expectations. The 162 hill town households studied by Lemuel Peet and his team averaged a net cash income of $601 per year, and about a third of the households reported as little as $400. Peet found that a shockingly low number: as he wrote, it was "almost impossible" for a family to make as little as that "and still maintain a satisfactory standard of living."[1] By comparison, a study of seventy-four "mostly prosperous" Vermont farm families two years later calculated a net cash income average of about $1,404—well over twice the average income of the hill farms. In that study, Agricultural Experiment Station home economist Marianne Muse acknowledged the difference between her findings and those of the Peet team, explaining that only a quarter of her sample households were the "poor type" of places that "Vermonters call 'hill farms.'"[2]

The eugenics investigators found similar stark contrasts in their own comparisons of income in Jamaica, Waitsfield, and Cornwall. There was poverty everywhere, to be sure. In each of the three towns, the team found several households that were "barely getting by," and a few who were desperately poor enough to be "on the town." One house in Cornwall was described as "hardly fit for human habitation," the children "dirty little ragamuffins."[3] Waitsfield's large farms employed tenant farmers who lived in tenement housing, in conditions the eugenics team found "unspeakable," in "worse conditions than most families in the city slums:" the children sick, the "baby hardly more than a skeleton."[4] In Jamaica, four people were "on the town:" all were sick or elderly people cared for in households that were reimbursed by the town. No households that included adults capable of work were living in quite such desperate circumstances as those few in Waitsfield and Cornwall.[5]

On average, though, cash incomes were markedly lower in Jamaica than in the other two towns (see Table 12).[6] Wages were around 20 percent lower than for those in Waitsfield and Cornwall. For farm incomes, the differences were more striking. Cornwall farmers who reported net cash income averaged a little over $1,000 a year, and Waitsfield farmers $1,602, exceeding even the Muse study's average. In contrast, the relatively few Jamaica farmers who reported their cash incomes as net figures did not even quite reach the $601 average the thirteen-town study had found in the hill towns. (Households that reported their income in gross figures showed similar differences.[7])

Table 12. Comparative Incomes, Net and Gross

	JAMAICA	CORNWALL*	WAITSFIELD
Total # households reporting income	85	81	94
Households reporting net farm income	22	45	53
Net farm income average	$573	$1,042	$1,602
Households reporting gross farm income	42	10	5
Gross farm income average	$771	$1,523	$,3832
Households reporting wages only	21	26	36
Wage average	$803	$1,020	$1,000

*Cornwall numbers are lower because fewer people answered this question.

To be sure, the household income figures collected by the eugenicists are missing several important pieces of the picture. For one thing, they recorded very little information about women's earnings. (The home economists at the Vermont Agricultural Experiment Station, in contrast, published a separate study in 1933 of "Cash Contribution to the Family Income Made by Vermont Farm Homemakers."[8]) The interviewers did take note of some women who held paying jobs. Most were teachers, nurses, and housekeepers, but they noted one woman who was operating a market garden; one who ran a farm and "worked out" logging; one who sold rugs; and even one who wrote for magazines. One teacher was reported to earn as much as $25 per week (more than the usual pay for men working on the bridges or roads), but most of the women's recorded incomes were very low. One housekeeper was reported to be working for just $7 a week. Another seems indeed to have been working only for room and board: the employer told Anna Rome that he "doesn't pay Mrs. [White] anything but that he occasionally gives her money when she needs it." (Mrs. White found it hard to get other kinds of work because she wished to keep her youngest child living with her. If she had taken a job away from home there would have been nobody to care for her child while she worked, so her "compensation" appears to have been room and board for the two of them.[9]) It seems unlikely that a systematic inclusion of women's earnings in the records would have done much to raise the town's average income.

Other Ways to Measure

Clearly, Jamaica households made do with significantly lower cash incomes than did those in Waitsfield and Cornwall, or in other more prosperous farming regions. But cash income was not the only way to measure "standard of living." That concept was originally intended to give the researcher better tools to measure the individual's access to valuable goods and important experiences beyond income alone.[10] As home economist Marianne Muse defined it in her first major study of Vermont farms' standards of living, the term referred to "the content of family living as measured by the amount, variety and quality of the goods consumed by the household in meeting the physical *and the psychic needs* of its members [emphasis mine]." Thus, "the cost of living in itself is an inadequate measure." A "fair picture" required supplementing income figures "with facts concerning the family, the home and the farm business."[11] With those facts, Muse attempted to assess the household members' access not only to food and shelter, but to nutritious diets, education, cultural experiences, and even such elusive goods as privacy and leisure time.

Marianne Muse was an important figure among Vermont agricultural professionals. She held a master's degree from Kansas State Agricultural College, one of the land grant institutions that pioneered in creating the discipline of home economics as an auxiliary of agricultural economics. Over the course of her long career with the Vermont Agricultural Experiment Station, Muse published the results of nineteen studies of household income, labor, and diet in rural households, some coauthored with her graduate students. Her work was characterized both by meticulous precision and by an abiding interest in lightening the workload of rural women. (She had a moment of fame in 1949, when her report on bedmaking techniques was picked up by the Associated Press and distributed widely.[12])

Muse's groundbreaking 1932 report on standards of living on Vermont farms might at first glance be taken for a eugenics survey. Like the team in Jamaica, Muse made use of elaborate questionnaires and personal visits with the women of the farm households. She collected data about household members' age, family size, educational attainments, and ethnic background as well as about income and expenses. And to be sure, her work sometimes exhibited a little of the professional arrogance so vividly on display in the eugenicists' field notes. But Muse had different priorities—perhaps one might even say different loyalties. It was her job to discover ways of enhancing the quality of rural life.

The eugenics team, of course, had a different agenda, and that difference was made plain when the two sets of experts addressed similar research questions. Both eugenicists and home economists, for example, studied housing conditions. In the Jamaica study, the eugenics team addressed that matter under the subheading "condition of home." In that space, the interviewer typically assessed the neatness of the rooms she saw; the physical condition of the interior (plaster falling, fresh paint); and the household's modern conveniences or lack of them. She might record a compliment—"living room is neat, attractive and cozy"—but she would more often note an observation like "things strewn about" or "bare furnishings."

Muse took her assessment of housing in a very different direction. Rather than focus on the housekeeper's care of the house, she considered how the house served the housekeeper. In one part of her analysis, Muse described the typically large old farmhouses most Vermont farmers lived in, judging that they did not, "as a rule, correspond to present-day needs." Built for much larger families, their size could be a burden to the women who had the responsibility for cleaning them. But Muse also appreciated that the average of two furnished rooms per person would provide "ample room for family life and for privacy"—one of the key "psychic" benefits she defined as part of the "standard of living."[13] Such an approach might have helped the eugenics team to gain a better understanding of living conditions in Jamaica, but it would have required a different attitude toward their subjects.

Beyond Cash

Most important in its implications for the hill farms studies, Muse and her colleagues conducted detailed investigations into the many non-cash forms of income that sustained most farms. Although non-cash income was never discussed in the Migration Study, many other farm surveys demonstrated that cash almost never provided more than two-thirds of the real income of rural people, and often not even half. In her own 1932 standard of living study, Muse found that the average farm produced the equivalent of $753 in non-cash income: $372 for food grown on the farm, $133 for firewood and ice, and $248 for the equivalent of house rent, amounting together to about 40 percent of household income.[14] (Home economists calculated the contribution of farm products consumed at home—food, wood, ice—at the price that would have been paid in local markets. They added a pre-set percentage of the value of the house to stand in for the rent the family would otherwise have had to pay.)

A larger 1936 study conducted by Muse collected data on the incomes of 960 rural Vermont households and found, similarly, that 42 percent of household income came in non-cash forms.[15] Nor does Vermont seem to have been unusual in this regard: an earlier nationwide survey of eleven different regions and nearly three thousand farms also reported the same average of 40 percent of income furnished from the farm itself, including two-thirds of the food.[16] And the thirteen-town study reported, similarly, that non-cash income made up about 46 percent of average incomes on the 162 hill farms examined in depth.[17] All these surveys found that the lower the household income, the greater the role of non-cash income. Households with the very lowest incomes might acquire as little as 10 percent of their income in the form of cash.

The eugenics team took no account of non-cash income for the residents of the three towns. If they had gathered that data, it would certainly have complicated their understanding of Jamaica households. Taking into account non-cash income, for example, might have lessened somewhat the disparity they found between Jamaica and the other two towns, since lower income farm households relied proportionately more on non-cash income than did higher income households. Jamaica's true household incomes would probably still have been significantly lower than other places, but not as low as they seemed to be. And clearly, including non-cash income would give a rather different impression of their experience of that standard of living.

Calculating the standard as "family living"—that is, what the household consumed, rather than what they earned—makes the difference plainer. The thirteen-town study found that the average yearly "family living" of its 162 sample farm households was $1095, including both what they bought with cash ($606 worth) and what they acquired through non-cash means—usually by producing it themselves ($489). Lemuel Peet (in a published article co-written with his mentor, agricultural economist Claud F. Clayton) acknowledged that the total "family living" figure was not as low as the team had expected, observing that "an average value of family living amounting to $1,095 suggests a quality of living that compares very favorably with that prevailing in mountain communities." In comparison, one study of Kentucky families found the total value of family living was less than $900.[18]

Taking into consideration the importance of non-cash income to rural households would not erase the basic fact that Jamaica households had lower incomes than many others in rural Vermont. (And of course, a favorable comparison with the famously destitute mountain communities of the Appalachians is not much to brag about.) But the failure to include non-cash income indicates a significant problem of perception. The eugenics team's blindness to the non-cash feature of rural life extended to the way they recorded farm production. As mentioned in chapter 4, they recorded none of the farm goods that households produced for their own use—nothing, for example, about the household's sources of heat, a matter of tremendous importance in the rural north. Equally important, they made no effort to determine the value of the home-produced butter, cream, milk, eggs, maple sugar, chicken, apples, beef, bacon, nuts, berries, jam, vegetables, venison, pickles, and potatoes that appeared on the household's table. Nothing about any of that—at a time when most people, whether rural or urban, spent well over a third of their incomes on food.

To underscore just how important an omission this was for the eugenicists, Muse and her colleagues reported in two studies of farm household nutrition that on most Vermont farms—even the poorest ones—diets were adequate nutritionally, with plenty of calcium and protein (they all had cows, after all), and enough of most other nutrients.[19] The 1930 study reported that some diets lacked sufficient iron, and the 1947 study suggested that some diets lacked enough vegetables, especially in spring when the stored vegetables and fruits had run out. But most farm households, including the poorest ones, ate three solid meals a day.

Moreover, in the 1947 study (which explored specific diets in detail), Muse and Johnston found relatively little difference between the poorest diets and the best. At all levels, the diet was heavy on bread and potatoes, but the mid-day meals always included some kind of meat, and the meals were supplemented with home-produced fruits, vegetables, and dairy products, and store-bought coffee, tea, sugar, and flour (and at higher levels, luxuries like orange juice and canned tuna).[20] Their only criticisms were those familiar ones nutritionists have so often expressed over the years: farm families should eat a lot less sugar and a lot more green vegetables. (Equally predictably, farm women countered that it was hard to get their kids—and their husbands—to eat greens.[21])

It is likely that the subjects of the second study—conducted over several years in the 1940s, in a prosperous farming region—ate somewhat better than did Jamaicans back in 1930. But perhaps the difference was not as great as one might imagine: Muse and Johnston suggested in the later study that many Vermont farm families did not change their diets a great deal after prosperity returned.[22] Most significantly, Muse and Johnston reported that "those with the best meals [still] raised much more of their own food than did those with the poorest meals."[23] Not recording—or perhaps not even noticing—these essential contributions of the farm to the household's security and well-being suggests that the eugenicists overlooked a critical element of how Jamaicans lived.

Modern Conveniences

Marianne Muse and her colleagues at the Agricultural Experiment Station viewed modern amenities as an essential element in determining the household's true standard of living. Indeed, they believed fervently in the power of new technologies. They put a great deal of effort into evaluating households' access to telephone, electricity, running water, and modern household conveniences (stoves, irons, washing machines, and sewing machines) that made work easier or life more comfortable for the women of the house. Muse's male colleagues, the agricultural economists in the Agricultural Experiment Station, kept a parallel account of farmers' automobiles and phones, as well as their access to electricity, machinery, and water in the barn.

Modern conveniences were often the most visible aspect of poverty and wealth in the countryside. The pattern of acquisition of such conveniences was quite spotty: access to some was determined almost solely by income, to others by proximity to a central place. The studies indicate, for example, that large numbers of rural Vermonters had access to some form of running water. Since springs and spring-fed streams and brooks were generally abundant in the hills, this was a relatively low-tech improvement for many farmers, who could often rig up gravity-fed systems to bring water to the barn and the house—usually just into the kitchen, but occasionally also into a separate bathroom. One 1922 study of farms in the central Vermont towns of Randolph and Royalton reported that

92 percent of farmers there had some form of running water. The 1932 Muse standard of living study reported that 95 percent of the farmers she surveyed had running water.[24] A slightly earlier Muse study of fifty households reported that almost half of the farms even had water piped into separate bathrooms.[25]

In contrast, electricity required extensive outside intervention and was much less evenly distributed. Just 9 percent of the thirteen-town study's hill farmers reported electric lighting, at a point when the overall state average was around 30 percent. But radios could be powered by battery if one was not connected to the grid. And the thirteen-town study indicates that almost two-thirds of hill farmers had automobiles, and an equal percentage had telephones.[26] Those numbers were very slightly higher than the statewide averages from the 1930 agricultural census. Perhaps the remoteness of the hill farms made such purchases a more urgent priority.

Unfortunately, the eugenics team did not keep a systematic account of which Jamaica residents had automobiles, telephone, running water, electricity, or other conveniences that might have contributed to a higher standard of living. They did occasionally comment on a household that possessed an admirably large collection of conveniences, or an unusually sad lack of them.[27] But for the great majority of households, nothing was recorded. If Jamaica households were like those in other hill towns, most would have had some form of running water; very few would have had electricity; and as many as two-thirds would have had a car and a telephone.

Jamaica villagers generally possessed more amenities than those out in the country, as one might imagine. A telephone company and an electric power plant were established downtown in 1901, though operation was sporadic in the early years. Electric streetlights appeared in the village as early as 1903.[28] By 1930, it is no surprise to find that the bank treasurer's house had all the modern conveniences: modern furnace, indoor plumbing, and electric lights. But one man with a relatively modest income was also adding a furnace and electricity to his house in the village when the eugenicists called on him.[29]

And even in the open countryside, household comforts varied. One couple who lived far out on Turkey Mountain Road surprised Marjorie Choate with their pleasant living room with its "rugs, window curtains, sofa cushions, canary birds, and furnace."[30] A modern furnace—one that was in the basement and sent heated air into a room or rooms of the house—was almost as much a social amenity as it was a physical comfort. Only about 17 percent of Marianne Muse's 1935 informants had one.[31] Traditionally, the wood cooking stove in the kitchen provided almost all the heat in farmhouses. Household members stuck close to the kitchen stove and closed off parts of the house to preserve heat. That was an important enough feature of rural life that the eugenicists frequently noted how many rooms were closed off on the chilly days when they visited. The effect of a furnace would be to make other parts of the house more comfortable and allow people to leave their more rustic kitchens to sit in genteel comfort in the front room—perhaps on a cushioned sofa alongside the canaries (birds who would be in serious danger of freezing to death without the furnace).

In another household out on the other end of town, in contrast, Elin Anderson noted the quite visible lack of modern conveniences. She recorded that at one "very shabby looking house" there was "wash out on front line." The wife was only too aware of that unfinished laundry—visibly "in need of mending," as Anderson reported from what must have been quite a close inspection. Low water levels had made it difficult for their gravity-fed running water to operate, and it had been impossible for Mrs. Rowland to do the laundry. She expressed an exasperated wish to "live in one good house before I die,"—one with "running water, electricity and some modern conveniences." Her household did possess some things that testified to a higher standard of living: a radio, a grand piano, and several other musical instruments. ("All the [members of my family] are musical," the mother explained.[32]) But they had very few labor-saving devices.

Hill farm households like this one were frequently pitied or condemned for their low standard of living. The lack of modern conveniences could place their poverty on display as clearly as the laundry on their front lawn. (All those musical instruments were probably not visible from the road.) And such pity was not necessarily misplaced: while Mrs. Rowland (or more likely one of her three music-loving sons) was carrying cans of water from the brook to the kitchen, Mr. Rowland was carrying water to the cows in the barn. Without electricity or running water, workdays were significantly harder and longer for everyone in the household. Still, judging from the testimony in the field notes, it was neither low income nor lack of modern conveniences that did the most damage to Jamaica's real standard of living.

Lost Opportunities

All three towns selected for the eugenicists' Migration Study had been declining in population since the mid-nineteenth century, but Jamaica was losing more than people: it was losing jobs. Cornwall was not much more than a "semblance of a village," as Anderson described it, an almost purely agricultural community adjoining the commercial center of Middlebury. Its residents had never relied much on town businesses or manufacturers for work opportunities. Waitsfield was a bustling commercial center of its own and a center for the surrounding communities, with a diverse economic base of service and retail establishments that provided work opportunities for members of farm households.

Towns like Jamaica, in contrast, relied to a much greater degree on the woodworking and lumbering jobs associated with the returning forests. By 1930, the sawmills and other woodworking enterprises in town were dwindling. And the losses were not confined to Jamaica. The thirteen-town team reported that in the period between 1880 and 1929, the number of woodworking establishments in the towns they studied had dropped from seventy-three to twenty.[33] Those jobs were essential to the survival strategies of Jamaica's poorest residents.[34] Without them, many people simply could not remain in town.

Perhaps an even more important disadvantage for Jamaica was its lack of a high school. The home economists at the Agricultural Experiment Station counted access to high school as an essential measure of a household's "standard of living." Technically, Jamaica students did have access to high schools in the nearby towns of Townshend and Londonderry, or to others farther afield, in Brattleboro or Saxton's River. Any student in a town without a high school could go to another town: their tuition would be paid by the town they lived in. But families had to be able to pay for travel, and for room and board if the student lived away from home. (The eminent Vermont politician George Aiken, for example, had taken the train eleven or twelve miles to and from his farm home in Putney to Brattleboro High School each day.) And of course, the family would have to do without the farm work or wages the student would otherwise have been contributing during those years. These were things some families in Jamaica could not afford.

One sixteen-year-old Jamaica resident told Anna Rome, for example, that she wanted to go to high school, but that since her mother's death she had been needed at home to keep house and care for her disabled sister.[35] Another family was managing to send their son to high school only with financial assistance from the wife's father to pay the bus fare, which cost "$2 ½ per week." (The father earned $4 a day as foreman of a railroad crew, but he had recently been hospitalized for several weeks "so that it took all they had.")[36] The oldest son in another household had managed to attend two years of high school, but "had to stop because his father was out of work and they couldn't . . . pay for his board and transportation."[37] High school was so important, indeed, that the *Brattleboro Reformer* listed it as one of the most prized but unexpected benefits brought to Jamaica by the "record-breaking" fern-picking year of 1923. The flow of money into town would mean that some "musty parlors" would be "cheered" by phonographs or radios that winter; some old houses would receive a much-needed coat of paint; and "some farm girls and boys can be sent to high school in Townshend, 13 miles away; for there is money to pay for the daily bus ride."[38]

Some young Jamaicans seemed to be facing a social barrier to attending high school as well as a financial one. In one household, a daughter told Elin Anderson she "would like to go to HS but cost of clothes and books is too great." Hattie was worried about paying for clothes suitable for high school, but she was also reluctant to leave home at all: in a long conversation, she confided to Anderson that she thought she would be willing to go as far as Londonderry (the closest high school from her side of Jamaica), but not as far as Boston: she thought she "would die of loneliness."[39] Several young people echoed that sentiment, telling interviewers that they were reluctant to go far from home for long periods of time, or that they had felt lonely or homesick when they tried it. One young woman had gotten through a year of high school "only after a great deal of trouble," then spent another year at a private school where "she did not get along very well:" the problem was that she was "homesick and so is back home again."[40]

Having a high school available in Jamaica would not, of course, have lightened Emma's family burdens, nor perhaps have prevented Hattie's shyness or worry about her clothes. But it would not at least have demanded a significant outlay of cash just to get there, and it might have fostered a more welcoming environment. The town of Cornwall had no academy of its own, but its village center was less than five miles from the high school in Middlebury, the commercial and educational center of the region. Ambitious students might have found it easier to attend high school in a town where their families often traveled for shopping or social events. In 1930, over a third of Cornwall's residents had attended some high school, according to the eugenicists' count. Waitsfield was an entirely different matter: six families reported to the eugenics researchers that they had pulled up roots and moved the whole household to Waitsfield specifically for their children to attend the well-regarded high school there. Over 40 percent of Waitsfield's residents attended at least some high school. In Jamaica, the figure was only 22 percent.[41]

All that might not have mattered so much in earlier years, when the district schools available in all towns provided an adequate education for most purposes.

But by 1930, high school was far more widely attended—both a clear marker of status and often a path to life-altering opportunities. The eugenicist investigators reported that education was an important factor in the outmigration of young people, and that attendance levels were higher among emigrants who left the three towns permanently. That was not quite correct, at least according to their count. In Waitsfield, the educational levels were a bit higher for those who left town (46 percent versus 42 percent), but in Cornwall, those who stayed home were *more* likely to attend high school than those who left town (36 percent versus 30 percent). Only in Jamaica did the numbers accord with the eugenicists' report: high school attendance was significantly more common among emigrants than among those who stayed at home (33 percent versus 22 percent). Simply living in Jamaica, unlike living in Waitsfield or Cornwall, was a barrier to education. Finding a way to get to high school, moreover, was likely to put a young person on the path to leaving home for good. The Peet study found that in the thirteen hill towns, "emigration . . . has been heavy and especially pronounced . . . in the case of sons who received advanced educational instruction."[42]

The "standard of living" in Jamaica may not have been quite as dismal as the eugenicists thought it was, but the community did face serious difficulties: limited opportunities for education, not enough jobs, and in many ways a less abundant life than might have been available elsewhere. Those constraints clearly played a key role in the decisions of many Jamaicans to leave town. One woman responded to questions about the emigration of her children with the same phrase repeated three times, once for each of her children: "no work here."[43] Several women reported they were contemplating moving "with or without husband" to a town where a son or daughter could attend high school.[44] Yet the field notes also suggest that at least some townspeople calculated the benefits and costs differently.

How Lower Standards of Living Might Be Better

Life in Jamaica was not always the worst option available to residents. Sometimes it might even have been the best. Under the right circumstances Jamaica's lower "standard of living" could actually make it a more appealing place for both newcomers and returning natives. Jamaica's declining population and shrinking economy offered some residents a chance to get a foothold that would not have been available to them elsewhere. One Jamaica man explained this to Anderson: "Admits that he would have gone elsewhere to buy a farm—but in good areas they are too expensive and 'poor people can't buy them.' So he bought a farm within his means here."[45] (This was the man whose wife was embarrassed by her laundry in front of the house.) Another told Anderson that he and his wife had been working at the New England Sanatorium, a Seventh Day Adventist hospital near Boston, but were both in "very ill health" and "felt that they would like to have a place of their own and be on a farm." They got a tip about Jamaica, probably from a co-religionist who was willing to offer a special deal: "A patient in the hospital knew of this farm in j; he was able to buy it with $100 deposit and so came here."[46]

That was one reason why agricultural experts worried that, in towns like Jamaica, once a farm had been "abandoned," it would not stay that way. If the price was low enough, the farm would be likely to attract someone else poor enough to buy it and try to make it work. (Once the Great Depression set in, that movement "back" to inexpensive land in places like Jamaica became epidemic—more about that in chapter 7.) Low prices were also one reason that retirement was a more common reason for moving to Jamaica than to the other towns.[47] In Waitsfield and Cornwall, retirees were generally older farmers from nearby who had transferred their farms to their children and moved to the village centers. But in Jamaica there were more retirees formerly unattached to town who were coming in specifically to retire there. One couple who had moved from Pennsylvania to Jamaica "to spend their old age and to be near daughter" explained to Anna Rome that they found Jamaica appealing because "it was cheaper to live here on their savings than in other places."[48] Another man confirmed that he had returned to town to retire in part because "his home had been in J," but also because he "found property cheaper here than elsewhere."[49]

Family Connections

In one other intriguing way, too, the situation in Jamaica might have operated to assist people without much in the way of money or connections. In all three of the towns the eugenicists studied, significant numbers of people mentioned some sort of family connection as a reason for their staying in town or returning to town. Some simply wanted to be near relatives; some married into town; some moved home to assist aging parents. The total number of people who mentioned family reasons was roughly similar among the three towns. But the type of family connection was surprisingly different in Jamaica (see Table 13). In Waitsfield, the majority of those who mentioned a family opportunity were men inheriting their father's farms. In Jamaica, only two out of fifteen were inheriting farms or businesses from fathers; instead, twelve out of the fifteen were in town because of help offered by the *wife's* relatives.

Table 13. Family Connections

	JAMAICA	CORNWALL	WAITSFIELD
Family ties, general	19	11	11
Help to family members	10	17	16
Help from family members	15	8	20
Inheritance from parents	2	4	11
Siblings	1	1	5
In-laws	12	3	3
Marriage	4	2	8
Other	0	0	4
Total family reasons	48	38	58
Family reasons as percentage of all reasons for being in town	17%	20%	21%

This information suggests that Jamaicans may have been operating with a different kind of social capital than was current in other places. In Waitsfield, perhaps because of the relatively successful and expensive farms at stake, the old patrilineal patterns seemed to be holding. The Waitsfield men who inherited farms often expressed a rather fatalistic attitude about the whole process: they had been born on the farm, but also into their role as farmer. Seven different men said "it seemed natural" or "it was natural" to take over their father's position. One said that he "fell into farm work." He explained that he "had dreamed of other things," but it had seemed "the most natural thing to do since he was brought up to it."[50] Four other residents used the phrase "it fell to my lot"—not exactly an expression of joy, but an acknowledgment of the mutual obligations by which farms had traditionally passed from one generation to the next. That system seemed not to be operating in Jamaica, where would-be farmers were more often in the position of calling on the help of relatives other than their parents: men set up partnerships with brothers or uncles; couples acquired or shared the wife's father's farm, the wife's brother's farm, or even the wife's grandmother's farm.[51]

One study of kinship patterns in Londonderry, the town bordering Jamaica on its north, describes a similar pattern there as a pragmatic response to the social strains of rapid population decline and the increasing difficulty of establishing new households. While land transfers in earlier times had been mostly from father to son, by the end of the nineteenth century, townspeople found it necessary to cast a wider kinship network to keep family assets intact. Thus, more land transactions took place horizontally, between brothers, sisters, and cousins—and between parents and daughters and sons-in-law.[52] The transfer of the Daniel Chase farm to daughter Fannie and her husband Arad Wood back in 1865 was one early example, brought on by the loss of two of Fannie's brothers to war and one to migration. It seems that Jamaicans, like Londonderry residents, were adapting to declining family size and narrowing options by casting a wider kinship network.

One young couple had cause to be grateful for this safety net, although they had to put up with a bit of local gossip about their situation. In 1930, Mr. and Mrs. Smith had just moved in with her grandmother, who had cared for Mrs. Smith as a child. Mrs. Smith told Anna Rome that she and her husband had bought the farm from her brother, who had lived with her grandmother on the farm before they moved in. But, as Rome reported, the "townspeople say the farm belonged to [the grandmother] and think it still does." Whether they formally owned it or not, it must have been a great relief for this couple to move to the farm. The Smiths had entrepreneurial ambitions. They had run a shop in one of the Jamaica villages for five years, and then "run an agency for packing ferns" for a company that distributed them to urban florists. But these promising enterprises both failed over the past decade. Now, Mr. Smith seemed excited about their new opportunity; he told Anna Rome that he had "always wanted a farm and so he is glad of the chance to farm and expects to make his entire living off the farm."[53]

Moving into Town

Most experts were convinced that hill people like Mr. Smith should move in precisely the opposite direction—go somewhere or do something else. They usually had in mind one of two alternatives: move to a town or city and get work for wages or find a way to emulate successful farmers in other parts of the state. The eugenics team, like many other experts, recommended that people in Jamaica be "encouraged" to migrate to a place where they could "better" themselves. But where was such a place?

From Waitsfield, high school degree in hand, some young men and women migrated to larger towns or cities where they were able to find white-collar jobs or attend college or vocational school. (The eugenics investigators counted seven Waitsfield residents who had attended college, and twenty-four who had been through vocational programs in teaching, agriculture, or business.) And a few Jamaica residents did follow the path through high school to white-collar jobs. One woman reported that her daughter had left home to live with an aunt in Boston, where she worked her way through a business college and landed a job as a bookkeeper: her proud mother said she was a "clever, ambitious girl and doing well."[54]

But that path was seldom possible for Jamaicans. (Only four of the residents counted by the eugenicists had attended vocational schools, and just two had attended college.) For most people who left town, the path led to factories, workshops, kitchens, and other farms—often not very far away. A third of all emigrants moved to the small towns that directly adjoined Jamaica, and nearly two thirds moved to other towns within Windham County. Clever, ambitious Ellen with the business degree had two brothers who had also left Jamaica, but they had not gone as far as Boston: one was a teamster in the adjoining town of Wardsboro, and one was a farmer in Dummerston, two towns away. Most of these kinds of relocations were motivated either by marriage or by a farm or job found in a neighboring community, often the same type of job or farm the migrant had held in Jamaica.[55] That was a move toward greater opportunity, to be sure, but not really what the eugenicists had in mind.

As the thirteen-town study confirmed, "only a few" of those who left were "engaged in pursuits that lead to great economic success or prominence."[56] Perhaps for this reason, when Jamaica residents did leave town and later return home, they more frequently described *worse* living and working conditions in other places, not better ones.

To Mills and Workshops

At first glance, it seems self-evident that almost any move away from a farm to a wage-earning job in a village or town would be a step up.[57] Workers in towns and villages clearly earned more than farmers did.[58] And in the industrial towns to which many rural Vermonters migrated for work, wages could be significantly higher than in the country. The lowest ranking employees in machine shops in southern New England—workers employed as "laborers"—averaged about $1,250 (figured as a year-round job of roughly fifty hours a week).[59] Perhaps Jamaica resident Mr. Prince made that much when he was working in a Connecticut foundry before returning to his farm. During his first year back home in Jamaica, he brought in only a little over a third of what he had earned in Connecticut.[60] Mr. Morris, similarly, had worked in machine shops in Massachusetts for years before he moved to his Jamaica farm. Wages for the skilled work he probably did might have been as high as $35 per week, around $1,750 calculated as a year-round job. The year the eugenicists visited, his farm brought in only about $700.[61] For wage earners in Jamaica, the average annual wage was just $803.[62]

At times when jobs were plentiful, workers in factories and workshops in downhill towns and cities almost certainly took in more cash than they would have earned in Jamaica. But working conditions in those factories and workshops could be brutal, even in comparison with the hard outdoor work that was most common in Jamaica. Mr. Kingston had worked for years as a machine tender in a paper mill in Claremont, New Hampshire, before his "health failed him in the paper mills and the doctor said he would have to be out in the country." (The treatment suggests he had a pulmonary disease.) He returned home to Jamaica, but as he explained to Anna Rome, that year he made a long and difficult recovery from appendicitis and was able to earn only around $500. New England paper mills reported wages of about $40 per week for machine tenders like Mr. Kingston, theoretically as much as $2000 per year for year-round work.[63]

Perhaps understandably, Mrs. Kingston told Anna Rome that she was impatient to return to New Hampshire. Rome recorded that Mrs. Kingston "Doesn't like it here and hopes her husband will be able to sell the place so they can leave.... thinks her husband would do much better elsewhere." (Rome rewarded her with a compliment: "Seems to have more ambition than some of the other women interviewed.") Mr. Kingston also expressed interest in returning to work in the mill "as he is well now." He must have meant he had recovered from the appendicitis: he said nothing about his recovery from the disease that had sent him back to the hills (asthma? pneumonia? tuberculosis?)—and nothing about the two fingers he had lost in the mill.[64]

Residents of all three towns studied by the eugenicists described bad working conditions: Mrs. Ames reported that her family had returned to Jamaica from Lisbon, New Hampshire, in part because the smoke from the factories made her children ill. A Cornwall resident reported he had damaged his health standing in water while he worked at the marble works in Proctor; and a Waitsfield man left a job with the gas works in Burlington because it was so "strenuous" it had damaged his health. But significantly more Jamaicans referred to their health as a reason for coming or returning to town.[65]

For some, too, the work was simply so unpleasant that it was not worth the higher wages. One man told Marjorie Choate that he had bought a farm in Jamaica because he was "sick of shop work" in the machine shops where he had worked all his life. Another reported that he had "quarreled at his place of employment—vowed he'd never go back—so came here." And Mrs. Prince had her own experience to relate: while her husband had been working in a foundry in Connecticut, she had worked in a mill. The work had been "hard and dull" to begin with, she recalled, but then on top of that, it had "become slack."[66]

Of course, work in Jamaica—whether in the farmhouse kitchen, in the field, on the roads, or in the woods—was no doubt also frequently "hard and dull." But most Jamaicans worked for wages for only a few months at a time, determined by the season and the availability of jobs. (The men, that is; Mrs. Prince's work on the farm in Jamaica was of course year-round.) Road work was done in the warmer months, opening sometime in April and closing in November. Most men who gave yearly estimates of their earnings to the eugenicists were referring to wages earned during that time. A few men reported logging or cutting cordwood in winter. Only three men who worked as laborers reported working for wages both in summer and in winter.[67] Most of these workers, after all, were also tending to animals and crops, cutting their own wood in winter, and probably tapping maple trees in spring and cutting hay in summer. Conditions in all these occupations were doubtless frequently hard, but they had the virtue at least of changing from season to season—perhaps a little less "dull."[68]

When Jobs Fail

Probably more important than the difficult working conditions in town was the fact that those better-paid jobs could—and frequently did—"become slack." Jamaica workers were accustomed to their own kind of "slack" seasons, but this situation was different: factories and workshops shut down or laid off their work forces in response to pressures far less predictable than the seasons. Anything from increased national competition to labor disputes to low water pressure might shut down a factory or workshop without warning. Both Mr. and Mrs. Ames, for example, had worked at the Jamaica Glove Company when it operated in town. (In fact, she had come to Jamaica with her mother specifically to work in that mill.) The business left Jamaica in 1913—lured by the Lisbon, New Hampshire, Board of Trade—and the family followed it over a hundred miles to work in Lisbon. They returned to Jamaica when the husband lost his job there during a cutback in 1920.[69]

And that was the biggest difference between Jamaica jobs and work in larger towns: in "slack seasons" in town, the workers were left without income, with rent still to pay and food to buy. In Jamaica, even households that depended mostly on wages almost always lived on farms or small properties that supplied a basic subsistence. Taking into account the food, shelter, and heat a farm provided, Jamaica-style working arrangements could be significantly safer than work in town. For some residents, a few months of city unemployment could make Jamaica look much more appealing—however meager the standard of living seemed to outside observers.[70]

And of course, even for a Jamaica resident who yearned to relocate to a more urban environment, the late autumn of 1930 was hardly an auspicious moment to begin. During the time the eugenicists were conducting interviews in the three towns, the national unemployment rate tripled. The next year it would come near to tripling again, and in 1933 it would reach its highest point with nearly 25 percent unemployed. In the quarries, textile mills, and machine shops of Vermont, layoffs and shutdowns were accompanied by dramatic cuts in pay for those who could still find work. In the major industrial centers, breadlines and municipal bankruptcies followed.

While they had no way to predict the depth of the disaster to come, the looming depression was very much on the minds of Jamaica informants that fall. The once-reliable work in the sawmills and woods was already increasingly hard to come by. Mr. Ames, who returned home with his family from Lisbon, had found no work at all in Jamaica since he had been home. His son had already been forced to leave high school, and it was still only the very beginning of the hard times to come. One man explained to Anderson that he liked farming very much and was "positive that he will be able to make it pay well." But more important than that, he thought, was that "at least he will be more sure of his daily bread than the laborers."[71]

That was how Mr. Andrews saw it, too. He had been working since he finished third grade: farming, "drawing logs," butchering, and "various kinds of work." He lost his job in Massachusetts when the establishment was "curtailing the help." Coming back to his farm this time, he was looking to provide his family with the basic security of food on the table and a roof over their heads. As he told Marjorie Choate, "the man who stays on the farm is much better off than the man who goes to the city."[72] Despite the constant outflow of migrants away from the hill towns, some Jamaicans still seem to have supported that proposition: most hill town residents who had access to a farm stayed there.

Among the adult sons of the 162 farm operators in the thirteen-town study, 64 percent still lived in the town of their birth or an adjoining town. Jamaica farm inheritors may have been even more likely to stay. According to the count of the eugenicists, only seven farm owners were among the hundreds of Jamaicans who left town between 1910 and 1930.[73] Once a young man started farming, Peet wrote, "he is apt to remain until old age or death." In the ten-year period of the thirteen-town study, the team found "only a few instances" where a farm changed ownership for any reason other than "death, old age, and infirmity." (And even those other reasons rarely "reflect dissatisfaction.")[74] There was little that could "induce" those who owned farms "to abandon houses, cleared land, woodland, public buildings, roads and other tangible evidences of wealth or assurances of security."[75] For men so situated, there was relatively little to tempt them to look for a higher standard of living in the towns around them.

Modernizing Farms

Leaving the farm for work in town, then, was not the clear path to success some experts thought it was. Simply acquiring a better farm was not a very realistic option for most people, either: Jamaica farmers were not likely to inherit one of those large and profitable farms more typical of Waitsfield and Cornwall, nor would they probably ever be able to buy such a farm. But what about the possibility of modernizing Jamaica's old-style farming practices? Might that generate a higher standard of living?

As discussed in chapter 4, there were numerous geographical and economic barriers in the way of such a project. Still, there were a few farmers in Jamaica who were operating more modern farms, and there were more who might have been able to do so. Judging from the field notes, the most successful modern farms in Jamaica—the ones the eugenicists labeled "progressive"—were those that were experimenting with ingenious combinations of crops, sometimes adding non-farm sidelines. Often, they were catering to select urban markets and to tourists. For example, Anna Rome called Mr. Livingston "one of the most outstanding farmers in town."[76] The Livingstons were the only family in town to sell strawberries, along with the more common crops: milk, eggs, beef, and maple syrup. Mrs. Livingston thought the other farmers in town could do all that, too, "if they took the time." Mr. Halvorson, the violin maker, sent his eggs on the express train to Boston; Marjorie Choate called the Halvorsons "one of the more enterprising families in J."[77] The Ryans, whom Rome described as "progressive people," ran a roadside stand and gas station, selling their produce, chickens, eggs, and maple syrup mostly to tourists passing through town.[78]

In later years, that strategy of marketing specialized products to high-end markets would gain the approval of Vermont's agricultural leaders. But in 1930, the consensus among experts was that the best path forward for Vermont farms was to transition to the production of fluid milk for the Boston and New York markets. The thirteen-town study concluded with the recommendation that, however many barriers there were to its successful implementation, "any sound farm organization" in the hill towns "must have dairying as its basis."[79] The research conducted by the agricultural economists at the Vermont Agricultural Experiment Station was thus focused almost exclusively on discovering the practices that would make dairy farming most profitable in Vermont.

The Experiment Station launched a series of dairy studies in 1921, making use of what one researcher called "the so-called survey method" pioneered at Cornell a few years earlier. Each year in the 1920s and 1930s, they published articles about these studies, reporting a wide array of results. Every farm needed higher quality stock to produce more milk per "unit," John Hitchcock's 1925 report on Randolph and Royalton explained.[80] But high-priced cows would produce more milk only if they were fed more purchased grain—of the kind and amount determined by E.W. Bell in his 1926 study of Cabot and Marshfield.[81] Investing in a very expensive high-quality bull might pay off, or it might turn out to be an unjustified "plunge," depending on the quality of one's already existing cow herd, reported a 1925 study.[82] A herd of around twenty-four cows would provide the best economies of scale, a 1926 study found.[83] But that size of herd would probably also require the farmer to buy a milking machine and hire a year-round worker, a 1929 study warned. As a result, the farmer would have to find an efficient year-round use for that worker—perhaps by milking in winter, when prices were reliably higher.[84] (Of course, that would require a reliable source of year-round labor. . . .)

To the experts, the path forward seemed clear, but it was a tortuously narrow one for most farmers. In the hills, the thirteen-town team calculated that to generate a reliable farm income, a household would need twenty-five dairy cows and sixty to seventy-five acres of crop land—a target that was utterly unreachable for the great majority of farmers. As Peet acknowledged, "not one farm in eight has sufficient size to permit of this organization."[85] But in fact, large numbers of farms in prosperous dairy farming areas would not have been able to meet that standard, either.[86] Judging by the areas studied by the Agricultural Experiment

Station, only the farms located on the very best dairy land in northern Vermont would have met Peet's requirement.[87]

And even in areas that were able to meet those standards of milk production, success came at a cost. The town of Enosburg, studied by agricultural economist Philip Hooker in 1924 and 1925, was in the flourishing dairy region of Franklin County. Enosburg farmers enjoyed a high standard of living: there was a high school in town; the farms were relatively well supplied with modern conveniences. Hooker praised the tendency of the farmers of Enosburg to share work and cooperate in the use of expensive equipment. But a closer look at his findings shows that cooperation in the use of equipment and shared tasks was not really a matter of choice in this town. Enosburg farmers operated much larger dairies than could usually be cared for by one operator, and they made use of very little hired labor. As Hooker explained, the town's average dairy size of twenty-two was "generally considered a sufficient number" to employ one man year-round "in addition to the operator."[88]

Enosburg farmers could not afford—or find—workers to sustain the size of their herds. (Hal Barron found, similarly, that back in the late nineteenth century it had been the limits on the availability of hired labor that prevented farmers from moving to dairy in the first place.[89]) On average Enosburg farms took on only about five months of hired labor per year, and they made up for the rest of the demand with extra unpaid labor from the members of the household. That extra work would have to be added to the farm household's already heavy workload. (Marianne Muse found in a study a few years later that Vermont farm women averaged summer workdays of 12.5 hours and their husbands 13.5 hours.[90]) And even that level of time commitment was not likely to bring in big profits, as E. W. Bell remarked in his 1929 study: "In order to farm successfully in Vermont ... one must watch the pennies."[91]

In fact, reading the dairy farm studies of the Vermont Agricultural Experiment Station, it is impossible to avoid concluding that the experts themselves found it difficult to chart a path toward reliable profits in dairy farming. Long before the Great Depression, prices for milk had been falling. A few years of high prices during World War I were followed by persistent losses in the 1920s. In the very first of the agricultural reports, published in 1925, researcher John Hitchcock wrote that the most "striking thing about the figures" he had calculated was "the meagreness of the financial return secured by the farmers of the region." His study focused on the farms of Randolph and Royalton, an area of modern dairy farms with many significant advantages over Jamaica, including good railroad connections and the rich valley lands of the White River. Hitchcock blamed the problem on a temporary downturn: "The years 1921–1924 have indeed been difficult ones for Vermont farmers." But he concluded that the "average farmer of the region little more than broke even when all expenses were taken into consideration."[92]

A few years of stable prices did arrive in the mid-1920s. But within a few more years—around the time the eugenicists were planning their Migration Study—another dramatic drop in dairy prices was hitting the state and the nation. Milk prices began to decline in early 1929, well before anything else went dramatically wrong with the national economy. By November, prices had dropped by almost a quarter. In just one year they dropped by half—far below the average farmer's cost of doing business. In short, during the period of transition to fluid milk as the mainstay of commercial farming in Vermont, there were very few years of smooth sailing.

Even in the good times, moreover, the agricultural economists repeatedly found that nearly all the farms they studied were technically losing money. For the sake of comparing farming with other businesses (and farming methods with one another), agricultural economists crafted complex accounting frameworks that would have been nearly incomprehensible to many farmers (or indeed to anybody outside their discipline). For ordinary purposes, a farm's profitability (whether it would "pay," as the old Board of Agriculture speakers would say) might be judged by the living provided directly to the household and by whether the cash income was sufficient to their needs. But agricultural economists needed a more detailed accounting system: they had to calculate what farm owners were earning, even if there was no actual cash flowing directly their way. The household must have a return on the capital invested in the farm, and it must be comparable to returns on other types of investments; the non-cash parts of the operation must be given market values. Profits must be calculated on a per-cow basis—counting for each "unit" all the inputs of labor, equipment, and feed, and balancing them against the price the farmer received from the sale of the milk from that particular animal.

Having created this system of measurement, however, the researchers found that almost nobody made a profit measured in that way. As researcher Harry P. Young found himself explaining in 1927, "'On the face of the returns the 186 farmers lost $3 per cow.'" That might look bad to a naïve viewer, Young acknowledged, but it "simply means" that the farmers "received less for their own labor and less for the roughage they raised and fed than they felt and said these were worth."[93] (This was the same H. P. Young who would soon become the chair of the Land Utilization Committee of the Vermont Commission on Country Life.)

Why Not to Modernize

That was in the good years. Once the Depression set in, the agricultural economists confronted a situation in which all their advice was useless: the bigger the farm, the greater the financial loss. S. W. Williams acknowledged ruefully that in the dire winter of 1931, "the average farmer in each district lacked roughly $400 of receiving any pay whatsoever for his year's work." If it had not been for the "living he had from the farm"—the house, the woods, the garden, the family cows and chickens—the household would have had nothing to live on. In Williams's understated phrasing, that year "the advantage of diversification was greater than usual."[94]

It is not too difficult to imagine that in these difficult years a well-informed Jamaica farmer might have been able to see through the haze of expert-generated statistics to a few pertinent facts. Let us imagine that this particular Jamaica farmer was unusually lucky: his land was indeed good enough and extensive enough to be adapted to modern dairy farming. Perhaps the barn was big, and weathertight enough to protect an investment in high-quality stock; maybe he had enough acreage to support those twenty-two cows, and the farm was located on a good road. Perhaps the household even had access to a little financial aid, from a relative or from the bank. Even with all these advantages, scaling up the farming enterprise would come at a significant cost. It would require extra hours of work from already heavily burdened workers. If the farmer was able to market milk in the more lucrative winter market, that would extend the work year-round. It would tie the farm operation to purchases of off-farm feed and fertilizer and commit it to the use of outside labor that would be hard to find and expensive to pay for. And it would make the entire household dependent on what surely must have seemed to be wildly unstable milk markets controlled somewhere far from Jamaica.

That was the situation that appears to have motivated some Jamaicans to take the "safety first" path described in chapter 4, even if they might have found it possible to venture farther into commercial farming. The patchwork of adaptations employed by many Jamaicans was no way to get rich, certainly. Diversified crops, wage work off the farm; household subsistence; ferns and trapping and "speculating" in cattle—these methods might provide a living, but they would probably not sustain a rising "standard of living." To limit the scale and ambition of a household's farming efforts in these ways would probably mean less cash and fewer opportunities. It would, in fact, require the household to accept that lower "standard of living" everyone associated with the hill farms. But it made its own kind of sense. Elin Anderson frequently wondered why some people in Jamaica seemed so "perfectly contented" with lives that seemed to her so poor. Perhaps some Jamaicans simply calculated that it was the best option available to them.

CHAPTER 6

Things Left Unsaid

Family, Community, Work

Incomes can be averaged; radios can be counted; nutritional intake can be calculated. Chapter 5 explored some of the results of those efforts to assess how Jamaicans lived. But some things cannot be quantified. For some residents, Jamaica's most important attractions were practical ones. It was a place to return when jobs disappeared or businesses failed or a place to stretch a limited income as far as it could go. But those were clearly not its only attractions, and perhaps, for many people, not the most important ones. One man made an indirect reference to those attractions when he told Anna Rome that he was glad to have exchanged his shop in Brattleboro for a farm in Jamaica: "There isn't as much chance to make money here," he said, "but there are other things that count more than money."[1]

Unfortunately, the man did not specify what those things were. That was typical: most Jamaicans did not say much to the eugenicists about such matters. Some of the most important questions, to be sure, were never asked. When the interviewers did ask, the responses they received (or at any rate, the responses they recorded) were usually brief and sometimes formulaic. (Of course, the brevity of these comments may have been the choice of the interviewers themselves, but they do not seem to have controlled the flow of information in quite that way. These notes, after all, were for their own use, not for public consumption. They seemed to record verbatim or close to verbatim when they could.) Yet even those fragmentary responses—*Wanted to be near family; It's home; Preferred farm life; More freedom here*—offer important clues. By combining those hints in the field notes with other sources, along with a little intuition and some guesswork, it is possible to get a sense of some of the intangible things that mattered most in Jamaica.

Family Ties

One late October day, Elin Anderson interviewed poor Mrs. Rowland, the woman whose water pressure was so low she was having trouble getting her laundry done. Anderson found the family interesting, and Mrs. Rowland "quite intelligent." Considering her laundry woes, it is not surprising that Mrs. Rowland told Anderson with some vehemence that she wanted to "get out of town." Anderson recorded her words directly: "I'd like to live in one good house before I die;" and "I'd like to live in a home with running water, electricity and some modern conveniences."

But in mid-interview, Mrs. Rowland made a quick retraction: "Doesn't really want to leave on second thought," Anderson wrote—because "relatives and people are all here 'and it's home.'" At that moment, Anderson's sympathy seems to have faded a little: she was still recording Mrs. Rowland's words verbatim, but she added a few barely noticeable embellishments. She placed quotation marks around Mrs. Rowland's whole sentence—and then a second set of quotation marks around that short, seemingly inconsequential phrase, "and it's home." She even placed a dismissive little "etc." on the end of Mrs. Rowland's final comment: "Not pleasant to feel strange among other people, etc."[2]

The eugenicists had little sympathy for such sentiments, associating them with the kind of dangerous passivity they diagnosed in Jamaica. Still, the few words Mrs. Rowland said about her family ties seem hardly worth Anderson's agitated punctuation. On paper, those words seem brief and unemotional, at least in comparison with her heartfelt desire for running water. Anderson probably understood, though, that Mrs. Rowland was referring to something more than personal relationships.[3] Five Jamaica households bore the name of Mrs. Rowland's family of origin, and that name went back for over a century in town. Her husband's lineage ran equally deep. And in this generation, her family and his were linked not only by their own marriage, but also by the marriage of Mr. Rowland's brother to Mrs. Rowland's twin sister. That sister was a widow now. She and her children were living on the home farm with Mrs. Rowland's parents—right next door to the farm Mrs. Rowland sometimes felt that she wanted to get away from. A complex network of kinship ties bound her to her community in multiple ways across both time and space. Perhaps she said so little about it, not because it was unimportant, but because there was too much to say.

Neighborhood Ties

Or perhaps she simply assumed everyone would know what she meant. Networks like Mrs. Rowland's were essential aspects of Jamaica life. They extended beyond close kinship ties to link entire neighborhoods. In that regard, Jamaica does not appear to have been unusual. In his study of the nearby tiny town of Landgrove (population in 1930, 104), historian Paul Searls disentangled what he called the "dizzying complexity" of its kinship network.[4] Anthropologist Robert Riley described the entire social structure of Londonderry (the town bordering Jamaica on the north), similarly, as "a series of linked, overlapping kinship groups."[5] That was still essentially true in Jamaica decades later.

The neighborhood around the old Chase/Wood farm, discussed in chapter 1, is one good example of how those linkages operated. Like the rest of Jamaica, this neighborhood had felt the impact of population loss over the decades: the Chase farm was now described as "the last occupied house on the road," and fewer families lived on that road now. But the connections among those who remained had been strengthened by generations of close interaction and frequent intermarriage, and by the assimilation of a constant trickle of newcomers.

The Bauer family, for example, were German immigrants who had migrated to Jamaica twenty years before the interviewers came to town; they were already deeply embedded in the neighborhood network. The daughter of the family had helped along their assimilation by marrying the boy next door, a young man with deep roots in town. (His family bore the name of one of the signers of Jamaica's original charter.) Back when young Arthur Rider was still living under his parents' roof, his family had taken in a boarder, Alex, whose father—also a German immigrant—owned a farm down the road. Perhaps Alex and Arthur talked over their strategy, because Alex also married the daughter of a neighbor: it was Alex and his wife who would take over the Chase/Wood farm, still known in their time as the "old Wood place."[6]

And that was not the end of it. Alex had a sister who left town and moved to Massachusetts, where she eventually married. But in 1910, with her husband in ill health, the pair moved back to her hometown and into the old neighborhood, a few doors down from her brother's farm. Mr. Leon died a few months before the eugenicist team came calling, but Mrs. Leon was still living on that same farm: now her daughter and son-in-law lived there, too—and he was yet another young man from next door (and who also bore the name of charter founders of the town).⁷ Those families clearly shared much more than a location.

As discussed in chapter 5, these kinds of tightly woven extended kinship connections could offer important financial benefits. But there was more to it than that: close family and neighborhood ties fostered shared economic strategies, but shared economic strategies also strengthened family and neighborhood ties. Underlying such arrangements was this neighborhood's shared preference for the subsistence strategy the eugenicists labeled "home use"—both a way of farming and a way of organizing family life. As was characteristic of such households, Mr. Bauer told Anna Rome that he kept no farm accounts, and that he really "had no idea" what he had made that year, though he made a guess at $500. And as was typical in such families, Mr. Bauer's two middle-aged unmarried sons had not moved away, but still lived at home, where they worked at all sorts of temporary, seasonal, and part-time jobs: "help father on the farm, work out quite a bit on other farms, do laboring jobs, trap, hunt and fern."⁸

Even when these two sons did leave home, they did so in a way that suggests a continuing commitment to a household economic strategy. The census reveals that the two brothers left home together sometime in the early 1930s and spent several years working as farm laborers on Staten Island.⁹ That migration does not appear to have been random: one of the brothers had been born on Staten Island. Perhaps it was an opportunity provided by extended family. After their father's death, the brothers returned home—together, again—to live with their mother on the home farm.

Meanwhile, on the farm next door, the sister of those brothers lived with her husband, actively replicating the same way of life for the next generation. The Riders, too, produced only "for their own use," they told Anna Rome. Their estimate of $500 for their yearly earnings was the same number her father gave, and it came from the same sort of patchwork of jobs her brothers reported: sugaring, ferning, and "working out." Just as her father had told the interviewer, Mrs. Rider also said that she had "no idea" what the family income really was. And like her brothers next door, her own young adult children—aged nineteen and eighteen—were still living at home when the eugenicists came to call. These two were the ones Anna Rome described as "staying at home and doing nothing, except for helping around the farm and house."

That was a family pattern the eugenicists particularly disliked. (Mrs. Rider may have suspected that Rome did not approve of her family, as Rome thought her even more difficult to interview than was her mother, who spoke little English: "she ... could not be drawn out, just seemed to tolerate the interview though she was not hostile."[10] The interviewers were convinced that adult children should follow their own aspirations, strive to maximize individual incomes and attain personal goals; normally that would be achieved by leaving home—and in Jamaica probably by leaving town as well.

And clearly, some people born in Jamaica agreed with them. Such close-knit neighborhoods and collective family arrangements were certainly not for everyone. It was a way of living that maximized basic security, but one that also made it quite difficult to pursue individual incomes and goals—and even personal preferences for marriage partners. It is easy to imagine that this way of doing things was what drove many people away from towns like Jamaica, toward more cosmopolitan places. Even the relatively crowded streets of small commercial centers like Manchester or Brattleboro might have provided a welcome taste of anonymity in comparison to the old neighborhood.

No Jamaicans said outright that they themselves preferred to be linked (or constrained) by such tight family bonds. Still, there is plenty of evidence in the notes that some of them found this kind of interdependence not just practical but comfortable, providing the security, continuity, and familiarity they expected and preferred. Like Mrs. Rowland, most residents spoke only in general terms: they preferred Jamaica because all their "people" were there; or they simply liked being near family. Some were a little more communicative, hinting at the familiarity and intimacy that cemented ties to family and community. A few proud mothers of stay-at-home children boasted to the interviewers that their children preferred to stay in town. Mrs. Tower's mother reported, for example, that her children "liked coming home on Sundays too much to go very far away."[11] 17-year-old Hattie Chipman told Anderson she would "die of loneliness" if she were to venture as far away as Boston.[12] Mr. Clay reported that he had been so homesick living in Brattleboro that he returned home "even though he knew he got better pay away."[13] Mrs. Randall's uncle was allowing her and her husband to rent their house from him, but that practical advantage was not the reason she gave Anna Rome for their return to Jamaica (from the town of Winhall, all of seven miles away): "Says this is the only place which seems like home to her."[14]

Social Ties

When Mrs. Randall said that she "felt homesick anywhere else," she may have had in mind not only the support networks offered by her immediate family, but the social connections she enjoyed in town. Jamaica maintained an active social scene, supported by a variety of leisure, sports, religious, and charitable societies. (Probably it could not be matched by the very small town of Winhall, population 229.)[15] By 1930, the thinning ranks of the old Civil War veterans' clubs had been gradually replaced by a new generation of social organizations: baseball, dancing, and automobiles were now playing more important roles. But those new town social organizations still operated alongside and within the networks created by families and neighborhoods.

Jamaica social life did not receive much attention in the field notes. But it did not go unrecorded. The *Brattleboro Daily Reformer*, the largest newspaper in the region, published frequent and detailed accounts from neighboring small-town correspondents about the goings-on in their towns. In the 1920s, the Jamaica correspondent reported on everything from private family events to club meetings to large-scale public performances. Births and deaths were recorded, of course, but so were residents' bouts of illness, new jobs, and visits to family members. In late 1930, the columns included accounts of several family Thanksgiving gatherings; a wedding; a woman who was "thoroughly surprised by a birthday party last Thursday evening;" a man who acquired a new radio; the number of hunters in town for the season; and a concert at the Jamaica Federated Church. They even reported the eugenicists' whereabouts: "Misses Roam [sic], Anderson and Choat [sic], who have been in town several weeks gathering statistics and records for the University of Vermont at Burlington, went home for Thanksgiving returning Sunday night."[16] Most important, the columns highlighted public performances and the doings of the formal organizations: the Benefit Club, the Women's Christian Temperance Union, the Masons, the Grange, the churches, the P.T.A.

The eugenicists, in contrast, collected information about Jamaica's social life only on one line of the questionnaire, labeled "Social Status." On most files they simply ranked the residents' "social status" as "excellent," "good," "average," or, infrequently, "poor." A simple system, but they seem to have found it unexpectedly complicated. The investigators seemed not at all sure what they meant by "social status"—or at least not what they wanted the category to convey about households in Jamaica. Ostensibly the idea had been to assess Jamaicans' degree of involvement in town events: the P.T.A., church services, charitable efforts. (More was thought to be better.) But the investigators frequently veered away from that original intention, instead filling in the blank with comments about wealth, education, occupation, or "respectability."

Anna Rome inadvertently drew attention to the confusion when she described talking with a woman whom she assumed was of a "good" social status because she was so well-dressed. Based on that perception, "one would think" that she would also be "active in social affairs"—but, oddly, she wasn't. "She isn't the least bit interested in social affairs and . . . she only belongs to the PTA." Rome ended up rating the family's social status "good," but she puzzled over it in her notes: "Seem to be a fairly good class of people. Live in a much nicer way and atmosphere than others of the laboring class making about the same amount of money."[17]

To make sense of these assessments, the interviewers piled on more and more explanatory phrases: "average according to the town," or "better than most of their class." Even that did not clear everything up. Anna Rome judged one family as "average, considering community," but she added two extra judgments: they were "respectable laborers," and they were "poor." Finally, she concluded with an entirely different kind of comment: "Wife comes from old J. family."[18] With that, Rome introduced yet another measure of status—and finally hit on one that was directly relevant to Jamaica social norms.

Old Families

Anthropologist Robert Riley's study of Jamaica's neighboring town of Londonderry suggests that throughout the nineteenth century, residents of that town mostly ignored the emerging modern social class categories so important in the growing cities. Instead, he found that "status, prestige, and access to positions of community power were largely derived from one's family and residential tradition."[19] A generation or two later, Jamaica's social organizations were still organized around that same "residential tradition"—that is, the long-standing presence of a family in town. Membership and leadership still reflected family identity more than modern markers of social class.

In a few social organizations, to be sure, important roles were filled by families who were distinguished by their wealth, education, or profession. As the *Reformer* reported, for example, the officers of the Jamaica Women's Christian Temperance Union in 1928 included the widow of a wealthy merchant; the widow of a former bank treasurer and their daughter; and the wife of the town doctor. But wealth and husband's occupational status were probably not the only things that made these women leaders. All the women listed as W.C.T.U. leaders were from families the eugenicists judged "good" or better ("excellent" or even one of the "outstanding families in town"). Not all had wealthy or professional husbands. One woman was from a family whose social status was described by Elin Anderson as "good—old farmers."[20]

Another woman on that committee was Mrs. Tower's sister-in-law. She and her husband were not wealthy, but the eugenicists perceived them as up-and-coming: Anderson described the husband as "one of the few really ambitious men in town," and the wife as "very bright and understanding."[21] Perhaps the interviewers themselves might have invited her to join a ladies' club for her personality alone. But that was probably not what earned her the place with the Jamaica W.C.T.U. Both she and her husband belonged to respectable families with long lineages in town, traceable back to the first generation of settlement.

By the same token, the financial worries of Mr. and Mrs. Tower did not prevent them from being members of several social organizations. Mrs. Tower herself was a member of the most important town women's organization, the Benefit Club, along with her husband's mother and his two sisters. Mr. Tower was a Mason, and Mrs. Tower was a member of its auxiliary, the Order of the Eastern Star. Both of Mrs. Tower's brothers were also Masons, and Mr. Tower—at the time of the interview an unemployed sawmill worker—had been master of the lodge the year before.[22] On that day when Elin Anderson questioned Mrs. Tower about how her family would survive the winter, she may not have realized that the couple was so deeply integrated into the town's social organizations. She struggled a bit to rank the family's social status, finally settling on "Good: wife is one of the old families," but she could not resist adding the word "Poor."[23]

There was some common ground between Jamaicans' measures of social status and those of the eugenicists. Obviously, the eugenicists valued "old families," too. And they were willing to grant Jamaica's old families some portion of the respect usually afforded to more exalted New England lineages in older and richer communities. Elin Anderson somewhat breathlessly described one Jamaica resident as a "direct descendant" of Vermont's greatest colonial hero, Ethan Allen.[24] No fewer than four households in town were reported to have Mayflower ancestry.[25] And Marjorie Choate even described one man—"a representative of one of the first settlers in J"—as having "somewhat of a stamp of aristocracy about him."[26] Mr. Mason was not a wealthy man: between his carpentry work and his potato crop he had made "very little more than enough to live on" that year. But it was not money that "stamped" him with elite status. Few in town could claim a better lineage than Mr. Mason of Masonville, a descendant of the founder of that village.

By one measure, similarly, Mrs. Rowland's family were merely "laborers," as Anderson labeled them. The unfinished laundry on the line confirmed both their poverty and their social class. But by another standard, as Anderson ultimately concluded, the family's social status was "fairly good," adding the explanatory phrase "old families."[27] Nothing is said in the field notes about this point, but it is worth pondering how the respect afforded to family name and long-standing residence might have influenced the attachment of some residents to home and "people." Unless a family name was very well known indeed—an Eliot from Harvard, a Roosevelt from New York—the status of "good family" was not very portable. "Poor" and "laboring class" status would follow migrants wherever they went. Family name and lineage immediately lost all their power to command respect when an individual left town. Even the 25-mile trip to Brattleboro—perhaps even the seven miles to Winhall—would render migrants anonymous. Maybe that is what Mrs. Rowland meant when she told Anderson it was "Not pleasant to live among strangers."

Cultural Ties: The Christmas Concert

Of course, social activities were not entirely monopolized by "old families," any more than by wealthy and distinguished ones. In real life, proximity, personality, and talent also counted for something. For example, a 1930 concert performed at the Jamaica Federated Church seemed almost to have made a point of drawing its singers from diverse ranks of Jamaica society.[28] The program was a collection of Christmas-themed texts adapted to fit well-known classical vocal selections for soloists and choir.[29] Among the soloists was a descendant of Ethan Allen, a baritone who sang a piece with music by Mendelssohn. (Anderson rated his social status "very good—leading people of town.") Mrs. Tower's bright young sister-in-law sang another Mendelssohn piece. The sister-in-law's brother (bearing their old family name) sang a tenor solo adapted from Wagner's "Song of the Evening Star." But there was also a duet by a hotel housekeeper (new in town, so no ranking) and a woman whose husband worked for the telephone company (ranked only "fair," with no explanation). And the organist was the wife of a man who ran a meat market. Her family was ranked "good;" she herself was "associated with the finest group in town."[30]

Perhaps it went without saying that the daughter of the former bank treasurer would be one of the eight soloists ("a fine old family, one of the best in town"). But so was a teenaged boy who had been forced by his family's poverty to drop out of high school that very year. The boy must have had a lovely tenor voice to overcome all the disadvantages he faced. French Canadian, impoverished, and recently arrived in town—his family had little status of any kind. They had acquired an eminently respectable English surname, but Anna Rome took pains to specify that it had originally been "some French name." She grudgingly acknowledged that the boy was "said to be good at school."[31] But everyone in town must have known that he was talented: his victory in a town-wide spelling match a few months earlier had been reported in the *Reformer*.[32]

Events like the Christmas concert suggest another way of interpreting what Jamaicans meant when they said, "people all here," or "it's home." Just as nobody directly said in an interview that they valued the respect and recognition that came from having an old family name, not a single Jamaican told the eugenicists that they were proud of being a big fish in a small pond. But the *Reformer* reported that the Christmas concert singers had an audience of 150 at the Jamaica Federated Church that December. Surely it was unlikely that any of these performers—even those who did have money or professional status—would have had the opportunity to perform in a program like this one for an audience in a larger town or city. And even if one of the singers had such an opportunity, it might not have meant as much as performing for an audience who knew she belonged to a respected old family—or that he could spell as well as sing.

Cultural Ties: The Pageant

As it happened, the single most ambitious public event of the decade in Jamaica was that year's Old Home Day, a celebration of the 150th anniversary of the town's founding. It took place just a few months before the eugenicists came to town, in August. The day-long festivities began at 9:00 a.m. with a band concert, followed by a morning baseball game (Jamaica versus West Dummerston) and then lunch. The front page of the *Reformer*'s morning edition reported that there were already over seven hundred people there for the morning events, with "automobiles parked everywhere."[33] After lunch, another band concert preceded the main event: a three-hour pageant re-enacting and celebrating the history of the town. The day continued with supper, followed by a three-act play and finally a dance. A reported twelve hundred people were on hand to witness the spectacle, despite an afternoon rain. Probably most of the audience came from out of town this time, because such an ambitious collective feat required the active participation and support of nearly everyone in Jamaica, a town of only 570 people in 1930. The *Reformer* counted 125 participants in the pageant alone, leaving for others the dinner and supper arrangements, the band concerts, the dance, the baseball game, and the acting, directing, costuming and set design of the play—to say nothing of traffic control, paying the bills, and all the other behind-the-scenes activities.

The pageant was written and directed by the daughter of a "well-known merchant," one of "the leading people in town."[34] The entire day was organized and financed by the Benefit Society, the women's club that functioned as a village improvement group. Among its other projects, the Benefit Society (including Mrs. Tower, her mother-in-law and two sisters-in-law) had paid for sidewalks in the village, benches at the local Salmon Hole swimming area, and repairs to the library. To fund such projects, they usually sponsored events that charged entrance fees: dances, plays, suppers, and the ordinary yearly Old Home Day celebrations. This day's events, however, were free: the pageant was an expenditure rather than a moneymaker, designed to showcase and celebrate all the talent and ability the town could marshal. It was almost as if it anticipated a visit from the eugenicists.

In fact, the program featured several of the men whose opinions would be most influential in shaping the eugenicists' perspective a few months later. The man who had those long conversations with Elin Anderson about his "planless" neighbors was the director of "Episode I, Father Time." The pastor who expressed his fear that Jamaica's unhealthy "contentment" might be rubbing off on him directed the episode about the War of 1812. The town physician played the role of the leader of the English soldiers in Jamaica's most famous (and only) colonial skirmish with Indigenous people. (He was the one who diagnosed Jamaicans as typical hill people, with "infinite contentment and no worry about the future."[35]) The enthusiastic commitment of these leading men to the pageant casts an interesting light on their conversations with the eugenicists. Even those who expressed serious criticisms of Jamaica social life could hardly have imagined that their remarks would later be used to bolster arguments for large-scale removal of their own neighbors from their farms.

Some particularly "good" village families appear to have had more than their share of starring roles in the pageant, but participants came from all over town.

One man from an outlying village, for example, was one of those old-fashioned hill farmers who had never left town.[36] (He was a cousin of the woman who told Anna Rome that Jamaica was the "only place that seemed like home to her.") He and his wife directed the third episode, "Early Settlers." They brought along a contingent of participants who traveled five miles down from their village to appear in an ox cart and present a song and dance "of pioneer time." Another family was part of the tiny Seventh Day Adventist community still centered in the mountain hamlet of Pikes Falls. They rode on the "Sugaring Float." (A photograph shows the couple and their children looking solemn and a little camera-shy, the wife turned sideways and gazing stoically away from the camera.[37])

In some ways, the most interesting event of the day was the play, although it came near the end of a very long day of entertainment. "Eureka Awakes" was a three-act play published in 1924, the story of a small town under the thumb of a domineering rich man; it featured an intrepid newspaper man, a local Yankee "character," and several plucky young women.[38] (The Benefit Society had already staged one of its author's plays, "Nora Wake Up," as a fundraiser in 1928.) In this case, several important roles did go to men and women who occupied high economic and professional stations in town. For example, the role of the villain, the most powerful man in town, was played by the son-in-law and heir of the merchant who might himself plausibly have been referred to as the "most powerful man" in Jamaica. According to Anderson, he was a genial and good-humored man; perhaps he found the role of the bombastic and domineering Mr. "Wargrim" an amusing departure from the personality his neighbors knew so well. Similarly, the role of Mrs. Nelson Dodd, "a would-be social power in Eureka, and the president of the Uplift Society," was played by the wife of the bank treasurer, who was in real life the president of Jamaica's own "Uplift Society,"–the Benefit Society.[39] Maybe she appreciated the play's caricature of "society ladies" like her. Perhaps it was a joke the town's leading citizens shared with the audience.

Elin Anderson wrote nothing about that. But she was impressed, reporting with something close to amazement that the "entire pageant was planned and conducted by the people of the town." One woman who had been coming to Jamaica for many years as a summer visitor was equally surprised, commenting to Marjorie Choate that "it seems as if it would be an impossible task to get up a pageant with the material they had to work with."[40] (Perhaps she was referring only to money, but the tone suggests she was thinking of the human "material.") Anderson thought it had been "one of the most outstanding performances given in years in the entire southeastern part of the State." That may sound like faint praise—"the entire southeastern part" of a very small state—but a look at the back pages of the *Reformer* reveals that the small towns in the area sponsored a great many plays, concerts, and other events, often with significant attendance from other parts of the region.[41] The pageant's success made a powerful case against Anderson's own not-yet-formulated theories about Jamaica's problems. If such a feat could be accomplished by a town with so few resources, even the most hardened eugenicist could not completely dismiss it. It did not alter her decision to recommend that Jamaicans be removed from the town, but she had to admit that the event was "proof of the fine ability and the capacity for cooperation that does exist among the people."[42]

Town Offices

Anderson did not seem to realize fully that Jamaicans, like most rural New Englanders, had frequent opportunities to demonstrate their "fine ability," though not usually in the presence of such large audiences. In addition to the charitable organizations and social clubs in town, the town government itself required some men to take on civic duties, and women were often enlisted as well. At the top, the town was administered by three selectmen. In 1930 there were sixteen other offices, including three auditors (who examined the town's financial records and reported to the town meeting on the town's fiscal condition); three listers (who assessed real estate values for taxation); and three school directors. All these offices were elected on a rotating three-year basis. The one-year positions—constable, town clerk, treasurer, town grand juror, legal agent, road commissioner, and the "agent to prosecute and defend suits"—were often also held by the same person over several years.

A relatively small group of men and women typically filled most of these offices. Just eight men held the position of selectman in the fifteen years between 1920 and 1935, for example: one for ten years, another for nine, and another for eight.[43] Six different people served as listers; fourteen as auditors; and nine as school directors, for a total of twenty-seven men and women in these major offices over fifteen years. This small group might be said to form a kind of governing class in Jamaica. But it was not made up of the same group of families who headed the social clubs and charitable groups. And there was little overlap among the different offices. A majority of the school directors, for example, were women (four of seven officers), and one woman served as a school director for the entire period. Most of the school directors had teaching experience or had been to college or teacher training programs. All were younger than most Jamaica leaders, perhaps reflecting the interests of parents with school-aged children.

Auditors were the most exclusive group. All were from "good" families, and most had some claim to "old family" status. Probably most important, auditors generally had some education that set them apart from other Jamaicans—at least high school, but often some college or business courses. At the same time, family still mattered: three of the seven auditors were closely related to each other. Two were a brother and sister, the children of the first treasurer of the bank. They appear to have inherited familiarity with bookkeeping and money matters: the brother succeeded his father as treasurer, and the sister clerked in the bank and in the post office the family also ran. A third auditor had married into the family by way of the brother's daughter.

The listers, in contrast, appear to have been chosen for very different qualities. Listers assessed the property value of all real estate to create the grand list for taxation; their work could have an immediate impact on almost anyone in town. Seven different listers were chosen over the fifteen-year period. One served for the entire fifteen years, two for ten and nine years each. Most came from families the eugenicists ranked as "good." But the three longest serving listers came from distinctly modest financial circumstances. All three were farmers. One was retired, living with a daughter and son-in-law who had serious financial troubles. Another was from the old Chase/Wood neighborhood; like his neighbors he farmed for his "own use," keeping no track of his expenses or income. The third lister reported that he earned "very little more than enough to live on." (As was common among those who did not keep books, he guessed it was about $500.)[44] It appears that in these important offices, voters preferred men with experience in old ways of farming and in making do with limited financial resources.[45]

The selectboard showed even more clearly the preferences of the townspeople for a specific kind of leader. Over the fifteen-year period from 1920 to 1935, there was consistent geographical diversity among the selectmen: one from the center village, one from a village on the west side of town, and one from a village on the east. When a selectman was replaced, it was always by someone from the same part of town.[46] And of the eight men who held the office of selectman between 1920 and 1935, only three had recognizably "old" family names. Although most selectmen were from households with a "good" social status as judged by the eugenicists, there were also men whose families did not rank so high. One was rated only "average" in social status, and two were specifically labeled "poor."[47]

This tendency would become even clearer in the early 1930s, perhaps influenced by the impact of the Depression. In 1934 and 1935, the select board included one man from an old and prosperous family. The second was described as "laboring class, poor, respectable." The third selectman was none other than John Tower, the husband of the woman who gave Elin Anderson so much trouble from her rocking chair. Mr. Tower was of a "good" family, but as Anderson recorded, he was also frequently out of work, and the family was clearly "poor."[48]

No resident told an interviewer that the reason they stayed in town was that being poor was not a barrier to civic participation. They talked so little about town politics, indeed, that it is doubtful the eugenicists even learned which of the interviewees held elected positions. But surely here, if nowhere else, was one reason to stay in Jamaica, where fellow residents would elect a man to one of the most responsible positions in town despite that terrible defect—poverty—that would weigh so heavily against him elsewhere.[49]

Working for the Town: Roads and Schools

Mrs. Tower does not seem to have told Elin Anderson that she came from an old family. (Probably someone else let Anderson know that.) She does not seem to have said anything about her active social life, either. Nor did she convey anything (or at least anything that Anderson recorded) about her family's tradition of officeholding. Perhaps the one man in town with the greatest practical power was the road commissioner. As manager of road upkeep and repairs, he had his own power base. His position required that he spend more money than most other officers in town, and he also earned a disproportionate amount of the money the town paid for road care. Between 1920 and 1931, the position of road commissioner was held first by Mrs. Tower's father; next by her oldest brother—the one Anderson called "one of the few really ambitious men in town"—and then by her second brother.[50] The two brothers were recorded to be the biggest earners of town funds in 1930. The Town Report listed the younger brother's total pay at $1112.35 in 1930. (The average gross income on Jamaica farms that year was $770.) The second largest earner was the older brother, with $728.78.[51]

Those incomes were probably not the product of corruption: the top four earners working for the town were all men who were able to supply their own trucks, which radically increased their earning power. And it appears to have been a requirement for the job that the commissioner take part in most of the actual work. (Two residents did complain to the eugenicists about the inequitable assignment of road work, but neither accused the commissioners of personal corruption.[52]) In that sense, the road commissioners and the other men who worked with their own teams of horses or trucks were members of a select group in town. Expert observers often had the impression that almost all the able-bodied men in the hill towns earned the bulk of their income by working on the town roads. But while it is true that many men did some work for the town, in 1930 two-thirds of all the workers paid by the town earned less than $100, and half of those earned under $10.[53] A couple of dozen men worked enough hours to make a significant part of their take-home pay, but only ten men earned over $200 in 1930.[54]

At its most casual level, town work might amount to little more than a single morning's wages. Over the course of the year, thirteen men worked removing "ice from road," for example, and most received $1.63 or $3.25, a half day's or a full day's work. That was the only pay five of the men would receive from the town for the entire year, and several would seem to have had little need for that small additional income.[55] Work like this may have been as much a matter of social obligation as wage-earning.

A similar arrangement seems to have held for the basic care of the local schools. In 1927, the South Hill school hired two sons of a local farmer as janitors. Their mother was also paid for cleaning the schoolhouse.[56] One man sold the school four and a half cords of wood, and another sawed the wood, receiving $5 for his labor. These were paying jobs, but they were also in some sense civic obligations. All the employees lived near the South Hill School, and probably all of them had personal and family connections to it. Several men, similarly, held positions as caretakers for one of the town cemeteries, typically the one nearest to them, where their own family members were buried. Such work constituted both paid labor and family obligation. In these ways, even work on the roads, the cemeteries, and the school buildings could be part of a social and civic network, further binding people to one another and to the place itself. None of that, of course, would be visible to the investigators. And no Jamaicans told them about it, either.

Dense interlocking family, social, and political systems operated in tandem to maintain the ties that bound some Jamaicans to their community. Some found their ties to family, neighborhood, and town difficult to break; others probably never thought of trying. Some preferred to live where they and their families were known, perhaps finding it difficult to imagine a good life in a place where their name and background meant nothing to other people. Some residents found good uses for talents a larger community might not have welcomed so warmly: they decorated floats, organized fundraisers, performed the music of Wagner. Some people, too, found meaningful roles in civic responsibilities, earning the respect and recognition of neighbors who trusted them with the difficult work of assessing property values, caring for the poor, overseeing the schools, and making frugal use of town resources. Some did no more than shovel snow (on the road to a neighbor's house), clean the schoolhouse (the one their children attended), or mow the cemetery (where their families were buried). Those tasks, too, were part of the town's dense network of connections.

And of course, some people escaped as quickly as they could for the bright lights of bigger towns.

Freedom To Stop Working

But there was one non-quantifiable, intangible part of life in Jamaica that brought some of those same people back from their migrations, and even tempted new migrants to the hills. That was the matter of workplace arrangements: not simply hours and wages, but the question of personal control over when to start; when to stop; how hard to work and how long. Unlike townspeople's experiences with family, social, and even civic life, this one related almost (though not quite) exclusively to the world of men. As with other personal matters, few Jamaicans spoke directly about their work preferences or the values that lay behind them. But in this case, the eugenicists helped to make things clear by recording their own outspoken distaste for the way things were done in town.

One subject in particular brought out Elin Anderson's most acerbic comments about men's work in Jamaica: deer season. It seemed to her to be a thinly veiled excuse for idleness. In her final report, she described it like this: "In nearly every home, at any time of day are to be found a group of men with old red caps pushed back on their heads, shotguns at their sides, seated around the kitchen stove—deer hunting." In another entry, Anderson implied with a pun that deer hunting season was a time for courtship: "Kitchen was filled with young people, composed of 'dear' hunters and the ... young people." In both cases, her grievance was clear: in deer season, "everyone feels free to stop work."[57]

Anderson was at least partly right about that. Her own notes suggest that late autumn was the beginning of a protracted period of relative leisure, when many men did not expect to work for wages. One man reported, for example, that he had "worked steadily except for November and December of last year." Another said that he had been laid off that November and did not "expect to have anything to do for at least 2 months and maybe all winter."[58] The interviewers noted that neither man seemed particularly concerned about his (temporary) joblessness. In fact, the typical work schedule for laborers and farmers in Jamaica varied a great deal by season, beyond even the normal variations in the farmer's year. Spring was a hectic time, when planting crops would have to fit in somehow between sugaring season and the start of the summer's wage and farm work. Long months of work on the roads or bridges, along with the usual work in the fields and woods, would come to an end as cold weather settled in, bringing with it periods of "slack" time when one might catch up on farm work, do repairs, or perhaps just sit around the stove "deer hunting." The beginning of the slower time of year coincided with deer season in November, followed by Thanksgiving and Christmas, traditionally times of harvest and leisure. Winter logging would begin after the New Year.

So, Anderson was certainly correct that hunting season was a time of relative leisure for many men. But hunting was a much more complicated matter than she understood. The field notes, with their focus on socializing around the stove, say nothing about any of the ways that hunting season was not just leisure, but work in its own right. They do not describe the experience of standing motionless in cold woods at dawn, slogging through briars and wetlands to track down wounded animals, or field dressing and carrying out a hundred-pound carcass. Nor is there any mention of the women's work at home that resulted from a successful hunt: the canning, brining, and sausage making that would be required to save the meat not used immediately. That was in keeping with the eugenicists' lack of interest in the non-cash work done by households for their own uses. But they also, inexplicably, neglected another type of work—paying work—associated with hunting season: many Jamaica men worked as guides, and many households provided room and board for out-of-town visitors. In 1930, the *Reformer* reported, for example, that "among the out-of-town hunters in Jamaica" were twenty-five men staying with five different families, along with "several at the hotel and many others."[59]

One account written by a native-born Jamaican suggests something of the interplay between work and leisure that actually constituted deer season. Muriel Follett's *New England Year* (1939) chronicles her life on a farm in Townshend, just over the border from East Jamaica. She described deer hunting season as a highlight of the year—so important that in her household Thanksgiving dinner was delayed until nighttime to accommodate the hunters. But her story complicates the perception of Elin Anderson that people felt "free to stop work" in hunting season. Follett's account includes the extra work of cooking for visiting hunters and of processing deer meat for her household. She reports rising before dawn each day to make breakfast and pack sandwiches for the forty or fifty men her husband guided into the woods.[60] But she also described her own enjoyment of the "tangy air, the heady, warming sunshine," and the "heightened senses of being in the woods."[61]

The eugenicists appear not to have encountered any female hunters, but Follett described her own hunting experience in some detail. For some rural women, hunting offered a rare form of leisure—the opportunity to spend time outside, away from their everyday, indoor household and farm work. Follett's account suggests that hunting season could provide the extra cash, food, and sense of accomplishment that rewarded good work, and at the same time, a refreshing break in routine. From her point of view, it was anything but idleness: "It's an exciting time, and hard, but [the men] love every minute of it. For some farmers, it's the only vacation."[62]

For some men, too, the work of hunting may also have been their most important contribution to the overlapping exchange networks of family and town—and simultaneously their greatest opportunity to shine in the community. In November 1930, while the eugenicists themselves were in town, the *Reformer* named seven Jamaicans who had already shot deer in the first two days of the hunting season. Several of those men were from the western, less civilized edge of town, where Jamaica met the more mountainous and depopulated town of Stratton. The list included a man from Pikes Falls whose wife told the eugenicists that very month that their income was "practically nothing." Another man on the list was one for whom the team recorded a particularly lurid family scandal. Still another lived in a shack with his brother, hunting and trapping for a living.[63] The social status of these men ranged from "average for Pikes Falls" to "very doubtful." Deer hunting was not itself disreputable, but it did offer scope for the skills of some men who might not otherwise have had many opportunities to gain recognition in the local newspaper.

Freedom To Work

Classifying hunting as leisure, the eugenicists diagnosed it as laziness. But even viewing it as work would not have raised its value in their eyes. In an insightful analysis of eugenics fieldwork, one scholar has pointed out that eugenicists often reacted with "extraordinary indignation" to the ways that "bad" families made a living—"berry picking, scavenging, itinerant farm work."[64] Work that was sporadic or unplanned; jobs that were unsupervised, seasonal or temporary; these types of work often generated alarm and disgust among those who studied the lives of the poor. That is how the eugenicists viewed much of Jamaica's patchwork of jobs—not just deer hunting, but also ferning, and even road work. The team was part of a long ongoing effort to encourage—or force—working people to work harder, longer, or in more regulated and controlled ways. That attitude was common among modernizing agricultural experts as well. It crept into casual references like the one found in L.C. Gray's influential agricultural economics textbook, which expressed the hope that the "easy-going farm life of the earlier stages of development" would soon "give way to a businesslike system of farming in which every action is measured by the dollar as a yardstick."[65]

Such a vision of productivity did not seem to appeal to many Jamaicans. But this, too, they rarely mentioned to the interviewers. A few residents commented that they disliked certain kinds of work: one was "sick of shop work;" another found work in the mills "hard and dull."[66] But as with other important matters, only a few spoke at any length. One told his story to Anna Rome. Mr. Dewey had been born in Jamaica, but lived most of his early life in Brooklyn, where his father had a wholesale maple sugar business. In 1900 he had returned to town to stay, "preferring country life to city life." Anna Rome jotted down his real motive almost as an afterthought: "Says there is more freedom here."[67] Another man explained to Elin Anderson in similar words why he had come home to take over his father's farm. He had learned, after working for other people for years, that "the work is harder and you never get your own freedom."[68]

What did these Jamaica men mean when they spoke of "freedom" in this way? A few hints from observers other than the eugenicists can shed a little light. One pair of new migrants to town offered their own perspective on some of the local connotations of the word. Helen and Scott Nearing came to Jamaica in 1932, not yet the iconic counterculture leaders they would later become. A notorious radical, Scott Nearing had been fired from several jobs, blackballed from teaching and writing, and ejected from both the Socialist and the Communist Party for his unwillingness to follow party lines. Cut off from most ways of making a living as the Depression deepened, the Nearings clearly valued Jamaica's low-priced land and cost of living. But they also hoped to find what Anderson aptly called "freedom from interference."

It is a little ironic, then, that the Nearings found their new neighbors a bit too committed to that kind of freedom. Years later, they recounted their experiences in *Living the Good Life*, a 1954 book that would later become a bible of the back-to-the-land movement when it was reprinted in 1970. Part of their agenda in the hills had been to create a rural hub for cooperative enterprises. But their attempts to convince their neighbors to join them in collective efforts of any kind had failed utterly. As they reported, all plans for systematic and organized work arrangements were resolutely rejected. "'Autonomous' is hardly the word" for hill farm households, they wrote: "'Sovereign' would be a more exact descriptive term."

According to the Nearings, Jamaicans were deeply protective of their control over their working lives. Tolerating very little work discipline, they "got up and went to work, or did not go to work, as a result of accident or whim." When they did decide to work, "they let inclination determine the object of their efforts." They could be turned from work by small distractions: "If the morning looked like rain or snow, they 'sat on their heels' in the local vernacular." Most frustrating of all, they prioritized social life over the job in hand: "If someone came along and wanted to visit, they would turn from almost any job and chat, sometimes for hours."[69]

Robert Frost, another migrant to northern New England, commented on the same phenomenon, though with greater sympathy, in a poem called "A Time to Talk." Frost was not writing specifically about Jamaica, but at the time this poem was published, he was coming to specialize in poetry based on themes intimately associated with northern New England's mountain landscapes and hill towns. (For over forty years, he was closely associated with Middlebury College's Bread Loaf writer's conference in the mountain town of Ripton, one of the thirteen hill towns studied by the Peet team in 1930.)

In this poem, Frost reveals his understanding of both old-fashioned farming and rural communities, carefully untangling the internal logic of a farmer's decision to stop work.

> When a friend calls to me from the road
> And slows his horse to a meaning walk,
> I don't stand still and look around

> On all the hills I haven't hoed,
> And shout from where I am, What is it?
> No, not as there is a time to talk.
> I thrust my hoe in the mellow ground,
> Blade-end up and five feet tall,
> And plod: I go up to the stone wall
> For a friendly visit.[70]

Laziness does not enter into this account: the farmer has already been working for some time. (The ground is "mellow" where he has been hoeing; it is cultivated deeply enough to push his hoe handle down into it.) The decision to stop work, as Frost explains, is based instead on two governing principles. There was a social obligation: when "there is a time to talk" it must be taken. And there was a fiercely guarded right to limit the demands of the always endless work—"all the hills I haven't hoed"—that stretched out before the farmer.

The Nearings were not without sympathy for Jamaica sensibilities. Turning their gaze for a moment back on themselves, they imagined how Jamaica farmers would view their own rather extreme devotion to organization as "self-imposed torture." "Those people work on a treadmill," they imagined their neighbors saying, "Why, they go on a schedule, like a train or a bus."[71] For many of the experts studying the hill towns, that was exactly the point. In a paper mill, or a machine shop, or on a more "dollar-measured" farm, their work lives would be more disciplined and organized. "Encouraging" people to leave Jamaica, after all, was not just a good solution because it would provide a measurable improvement in their "standard of living." Even more important was something unquantifiable: a more disciplined and efficient approach to work, with fewer times to stop and talk—whether gathered around the wood stove "deer hunting" or putting down your hoe and meeting your neighbor. Reading between the lines, it appears that those were precisely the opportunities that kept some people in Jamaica.

Some people had clearly migrated to Jamaica for the same reasons. Miss Jensen, for example, had a lot of experience with treadmill work, as she told Elin Anderson. Born in Finland, she had worked hard most of her life, first as a domestic servant and then in factories, which had at least granted her "a little more independence." She and her sister had discovered Jamaica on a summer visit. They "wanted to be free and out in the country," and in 1923, they came to live there. At the time of the visit from Anderson, the sister had recently died, leaving Jensen to operate an 85-acre farm on her own, growing potatoes and hay on a "slope of land" with a "lovely view of the mountains." Anderson described her as a "respected Old Maid [her capitalization]," a "short wiry woman dressed in knickers and sweater." These days, Jensen made a living in classic Jamaica style: tapping maple trees, ferning, and helping her neighbors with logging. ("It isn't strength required so much as skill in handling the horse.") She reported a total gross income of just $350 that year; she could have made more, she told Anderson, but it "just isn't worth it." Now, living in a remodeled chicken house, Jensen seemed to Anderson to be "perfectly at ease." She finally had "the one thing" she and her sister "had wanted all these years"—"to be free and able to do as they pleased."[72]

Newcomers like Miss Jensen were becoming more and more common in Jamaica in the 1920s. And it would not be long before Jamaica and the other hill towns would encounter a great wave of such migrants, who, like Helen and Scott Nearing, were fleeing an economic crisis that threatened to destroy all individual plans for autonomy, work, or leisure. The Depression-era migrants would bring a new mix of ideas about all those matters. As things turned out, many would come to Vermont believing that the solutions to their problems were to be found in the hill towns' old ways of doing things.

CHAPTER 7

The Great Depression

How the Problem Became a (Temporary) Solution

Back in 1929, over a hundred members of the Vermont Commission on Country Life came together to hear progress reports from its committees.[1] The final speaker of the day was Arthur Peach, professor of English at the Military College of the State of Vermont. By late afternoon, it had become clear that some of the other committees were formulating big plans to tackle the problem of the hill farms by getting rid of them altogether—converting the land to more profitable and constructive uses as forests, parks, and tourist destinations. Professor Peach spoke against those plans. As befitted his role as chair of the Committee on Traditions and Ideals, he reminded his listeners of the traditional values associated with the hill farms: "independence of spirit, love of the hills, some sense of freedom that far hill vistas give." But in retrospect, Peach's most memorable argument was not about "traditions and ideals" at all. It was not only because they were expressions of old-fashioned values that the hill farms should be preserved, Peach pointed out. They were something more important than that: "home[s] from which a man can neither be starved nor frozen as long as a crop will grow and wood burn."[2] Indeed, less than three weeks after that meeting, the Black Friday stock market crash would tip off the cascade of disasters that became the Great Depression. "A home from which a man can neither be starved or frozen" was about to become a much scarcer commodity.

The Hill Spirit

The agenda of the Country Life Commission had taken root in the business-friendly, rapidly urbanizing 1920s. (A later critic referred to it as "the high unpleasant noon of Coolidge prosperity."[3]) In those years, many educated and progressive Vermonters were certain that, one way or another, the hill farms were destined to disappear: the only question was how to make good use of that inevitable change. But Peach was not the only one who expressed misgivings about the plans the Commission contemplated. To be sure, progressive men and women feared that the poverty and backwardness of the hills would hold back the rest of the state in its efforts to build a modern economy. But they sometimes worried, too, that the modernization of the state would come at a high cost.

A vivid depiction of that ambivalence appeared in the pages of the literary magazine *Driftwind*, launched in 1926 by Universalist minister-turned-shopkeeper, printer, and editor Walter Coates. *Driftwind* was devoted mostly to avant-garde poetry, and its contributors were generally left leaning: strongly anti-war, defenders of Socialist Eugene Debs' right to free speech, and supporters of child labor laws. Culturally the writers were given to strong statements against materialism and commercialism. But Coates also identified his magazine as a voice for Vermont—and specifically a voice for the hills of Vermont. The masthead of the magazine read "*Driftwind* from the North Hills," followed with the motto "The Hill Spirit, Today and Yesterday."

In 1927, *Driftwind* writers engaged in an extended debate over a new state campaign promoting tourism, a critical part of most plans for modernizing the state. Journalist and historian Walter Crockett broached the subject in a pro-tourism essay called "Will Vermont Look Forward?" (Crockett had been director of the Vermont State Publicity Bureau in the 1910s and was an enthusiastic supporter of the promotion of tourism. He was soon to be an important member of the Country Life Commission.) The state was in grave peril, Crockett wrote: "There is nothing to be gained by concealing or ignoring the facts." There were only two options: Vermont "settles down in apathetic acquiescence" to "a steadily diminished population in the hill towns and the loss of her brightest and best young men and women"—or Vermonters would "adapt themselves to changing conditions" and engage fully with the tourist industry, cooperating to "bring thousands of city dwellers" to the hill farms, "where summer houses may be established."[4]

Apparently anticipating outraged responses to Crockett's essay, editor Coates attempted to soften the tone of the debate in advance, requesting "candid, sane, constructive discussion." The responses he received were candid, at any rate. One writer declared that the idea of "selling Vermont beauty"—the slogan of the campaign—made him "nauseated." The pro-tourism agenda was the creation of "a species of flotsam, parasitic-Babbitt group," and the slogan "reeks particularly un-Vermonterish and painfully objectionable to me." If this is progress, he wrote, "give me something else."[5] A second writer agreed, adding that it was those who had been "afflicted with the fever of modern life, in its rankest materialism and commercialism," who "want to turn Vermont into another land boom inflation of the Florida or New Jersey type."[6]

Another supporter of tourism responded by re-stating the gravity of the situation: "We cannot claim that nothing is the matter with us." Not only was the state's population "stationary," but its "quality has deteriorated more than the quantity." (The pro-tourism writers often subscribed to the notion of declining population "quality.") This commentator was Paul Prentiss Jones, who operated a farm in Windham, just north of Jamaica. But he was also a well-known columnist for the *Brattleboro Reformer*, writing under the name of "Rustic." As he saw it, "Agriculture alone ... will never enable us to advance ... but scenery is one of our greatest resources. Why is it so despicable to let people know about it and enjoy it?"[7]

Editor Coates attempted to placate both sides by re-phrasing the language of the slogan. To those who found "selling Vermont beauty" to be an offensive phrase, he explained that it was a modern metaphor: to "sell" meant to "awaken interest in, admiration for, attachment to, constructive effort on behalf of, Vermont, her soil, her woods, her lakes and streams." In the end, he gave up on persuasion, ending the debate by announcing that "*Driftwind* has enlisted!" as he joined the Vermont Commission on Country Life. (Coates actually joined Professor Peach's Committee on Traditions and Ideals, which perhaps ironically included members from both pro- and anti-tourism camps, and often expressed not only ambivalence but directly contradictory opinions about all the new plans for improving Vermont.) After that (apart from one author's tongue-in-cheek and increasingly extravagant calls for Vermont to secede from the union) the magazine went back to publishing poetry.[8]

Regionalist Stirrings

Driftwind was not the only setting where faith in regional "traditions and ideals" jostled with a modernizing agenda for the future. In the magazine, at least, the tensions were articulated by different people. In the case of Dorothy Canfield

Fisher, all the contradictory positions coexisted uncomfortably within a single person. Fisher was a best-selling novelist with an international following, well-traveled and multilingual; she was among the extremely small number of women in her generation who earned doctoral degrees; her twenty-five years as an editor of the Book of the Month Club made her a well-known public intellectual on a national stage. She was also a lifelong resident of Arlington, Vermont, less than twenty miles from Jamaica as the crow flies west over the mountains.

As a public figure, Fisher played an important role in charting the course taken by Vermont's leading modernizers. She was an active member of the Country Life Commission and supported most of its policies, including the state acquisition of hill farms for conversion to forests.[9] For decades, she was a tireless and creative advocate of the kinds of tourism she thought would help to modernize the state without destroying its special characteristics. And yet, as a writer and a speaker, Fisher dedicated most of her life to an exposition of those special characteristics, always associated with the traditional rural life of the hills.

As early as 1917, Fisher published what remains her best-known work, the children's story *Understood Betsy*. It was there she first made her case that rural Vermont households, for all their poverty and isolation, provided the healthiest and most morally elevated environment for living.[10] The setting is the generations-old "Putney" farm, where plain living, prickly independence, and Yankee humor combine to make a perfect home for the orphaned city girl Betsy. Over the course of the story, Betsy's new environment teaches a series of lessons that transform her from a timid, fragile, and self-absorbed child into a self-reliant and upright member of her rural community.

And *Understood Betsy* was just the beginning. Fisher's distinctive ideas about Vermont remained at the center of her philosophy of life, expressed for more than four decades in novels, short stories, essays, articles, and speeches.[11] Throughout the 1920s, she continued to expand her vision of rural Vermont's ethical and cultural superiority. In a 1922 article in *The Nation* titled "Vermont, Our Poor Little Rich State," she made the claim that the people of Vermont (prototypically rural, as Fisher always described them) were chiefly distinguished by their freedom from the fear and envy that haunted other Americans. A Vermonter had no fear of being poor "because he is poor already and has been for a hundred and fifty years, and it hasn't hurt him a bit." Vermonters did not envy one another, she proposed, because the only basis of social status in rural Vermont communities was doing one's share of productive work—"to raise food, or grind corn, or make shoes . . . or teach children." The "Vermont personality," she concluded, was "tinctured to its last fiber by an unenvious satisfaction with plain ways."[12]

By the late 1920s, Fisher's laudatory depictions were beginning to be joined by the work of other Vermont writers exploring similar themes. In 1928, just as *Driftwind* was engaged in its tourism debate and the Vermont Commission on Country Life was in its initial planning stages, the poet and drugstore owner Walter Hard published his first book of verse. *Some Vermonters* was a collection of un-metered, non-rhyming poems designed to capture the distinctive speech and thought patterns of hill farmers. Hard himself lived in Manchester, a bustling commercial and resort center in the valley of the Battenkill River on the west side of the Green Mountains. But the mountains rise sharply on the east side of Manchester, and the ridgeline runs through Winhall, the one town that separates Manchester from Jamaica.

Hard's classically wry, understated Yankee figures made his work popular both within and outside the state. (In 1939 his fifth collection *Vermont Valley* garnered a positive *New York Times* review which called the poems "pungent" and full of wit.[13]) The poems were seldom too flattering or sentimental; Hard was candid about the often harsh and confining living conditions in the hills. He did not debate the policy alternatives the Country Life Commission was formulating for the hill towns. But he did make it clear that he admired the beauty and integrity of the difficult and increasingly threatened way of life in the hills. The experience of the Depression would deepen his appreciation.

In 1932, Hard turned to prose to express his growing conviction that the way of life on Vermont's hill farms had something more valuable to offer than a nostalgic glimpse at old ways of living. His wife later recalled that he had been inspired by a conversation with a Manchester summer resident, the publisher John Farrar, who asked why the Depression was not causing in Vermont "'the . . . desperation and despair' . . . he had experienced in other states."[14] Hard's response to that question became his magazine essay: "Vermont—A Way of Life."[15] In a sidebar to the essay, the magazine's editor predicted that readers in that difficult time would take an unusually strong personal interest in how Vermonters were weathering the Depression: "In good times, the people of Vermont made a living—and were rather looked down upon as unambitious." Now that the nation had fallen on hard times, Vermonters still do no more than "make that same living," but now "millions envy them."

Hard's claim was a bit extravagant: he proposed that rural Vermont society produced people who were uniquely qualified to cope with the Depression.

From early childhood, rural Vermonters were taught the old-fashioned values of thrift and frugality, along with the practical skills necessary to make do with little. Taught to place the highest value on self-reliance and household self-sufficiency, they encountered no social pressure to live beyond their means. No one mocked them for their poverty; no false shame prevented them from living modestly. A Vermont family might not enjoy getting along without a car, but their self-respect would not suffer without one: on the contrary, they would be ashamed to have their neighbors know that they were "rid[ing] around in something they could not afford."[16]

World Turned Upside Down

If this way of imagining rural culture seems a little familiar, perhaps it is because it bears a rather striking resemblance to Elin Anderson's 1930 analysis of what was *wrong* with the hill farms. The idealized depictions of Vermont by Walter Hard, Dorothy Canfield Fisher, and increasing numbers of other authors, are almost like the eugenicists' reports turned morally upside down: the same characters appear, but they mean something completely different. One small but startlingly exact parallel stands out from a Jamaica interview. Elin Anderson recorded that the eight-year-old daughter of the Tower household walked to school and back on her own, "even in winter." Anderson disapproved, evidently thinking of the trek as a threat to the child's physical and mental health. But that exact story appears in *Understood Betsy*, where a walk to school—alone, in winter—is portrayed as the beginning of fictional nine-year-old Betsy's journey to independence and robust health. It is almost as if Anderson had rewritten *Understood Betsy* to include harsh, careless relatives who mistreated little Betsy in place of the wise, kind, Yankee relatives Fisher had imagined.

Similarly, Walter Hard's claim that rural Vermonters were not ashamed of their poverty offers another mirror image of Elin Anderson's analysis of Jamaica. Anderson's comments seemed at times to imply that if Jamaicans were not ashamed of their poverty, they probably should have been. A healthy environment should encourage striving, competition, and envious glances at what other people had. Thrift, contentment, "satisfaction with plain ways"—these were all very well; but a little envy and shame might provide just the incentive Anderson thought was lacking in Jamaica. A social environment that encouraged residents to be contented with poverty was a recipe for genetic decline.

Of course, neither Hard nor Fisher was a disinterested observer. In some ways, they may not have understood their neighbors much better than did the three eugenicists. Walter Hard later explained ruefully that he did not know much about the rural self-sufficiency he praised: "I'll have to confess that I never have had a successful garden." Nevertheless, he wrote, his observation of other people's self-sufficiency made him believe that "given some land, some chickens, a pig and a cow, I could feed my family."[17] Fisher, for her part, clearly exaggerated the independence of Vermonters, writing in the 1930s that she and her neighbors literally could not comprehend the lives of "poor modern urban wage-earners" who were "gripped fast (so we understand) in the rigid, impersonal framework of a society organized uniquely around money."[18] Yet the 1940 census schedules show that over 40 percent of the 628 wage earners in Arlington—the town she lived in for fifty years—worked in factories.

Still, just two years into the Depression, Anderson's perspective was beginning to lose ground in Vermont, as the old-fashioned virtues—self-reliance, frugality, contentment—were looking increasingly attractive. And in that spirit, Walter Hard upended the meanings assigned to the hill farmers. Rather than imagining the farmers in places like Jamaica as problems to be solved by experts, Hard now presented their way of life as a solution to the daunting problems that faced the state and the entire nation. Far from being a burden to their fellow citizens, he argued, hill farmers were the guardians of the state's most valuable traditions. No longer at the margins of Vermont society, they were now central to its identity: "That is habit. It is also tradition. It is Vermont."[19]

Vermont Becomes a Way of Life

By the mid-1930s, this way of thinking about tradition, hill farms, and Vermont was acquiring a new and appreciative audience, as a flood of exurbanites began to find their way into the state. Suddenly, Vermont was awash in writers, painters, architects, and musicians, who all seemed to be leaving cities for farms. Low real estate prices certainly played a role, and so, of course, did the collapse of all kinds of work opportunities in the cities. But there was also something happening to the popular image of the state. And these Depression-era immigrants were not simply responding to that emerging image: almost as soon as they set foot in the state, they began to contribute to its development.

The writer Elliott Merrick, for example, moved with his family to Vermont in 1933, fleeing the Depression for an "abandoned" farm. Just one year later, he published *From This Hill Look Down*, a fictionalized account of his own "back to the land" experience in Vermont. Merrick was not just a migrant to the state; he was something like a convert. The very title of his book, taken from the first poem in Walter Hard's 1933 book *A Mountain Township*, conspicuously announced his loyalty to the new vision of Vermont. And there were dozens of writers like him. By 1938, Harvard historian Crane Brinton was exclaiming over the work of "literary outlanders" like Merrick, who were relocating to Vermont and flooding the market with books and articles about it.[20]

Some of the books on Brinton's list were genteel country life "I-bought-a-barn" books like those of Frederick Van de Water, who chronicled his journey from Manhattan to Dummerston, Vermont (two towns downriver from Jamaica), in *A Home in the Country* (1937). Others, like Merrick's, took a more serious tone, coming to terms with the possibility that their generation might be living through the permanent breakdown of centralized industrial capitalism. Vermont's traditional rural society now seemed to offer a lifeline to people fleeing a failing system. Like the farmer in Walter Hard's poem "The Mountain Farm," who refuses an outlandish sum of money from a summer visitor for his hardscrabble farm, the prototypical rural Vermonter these writers imagined was a new kind of role model. He had refused to worship growth, prosperity, and mobility as others had done during the 1920s, and now it seemed he had been proven right.

Elliott Merrick's message was stark: at a time when all systems were failing, it was essential to be able to provide one's own food and shelter. Summing up that message in his second Vermont book, *Green Mountain Farm*, he described a shared Christmas dinner with a neighbor. She apologized in advance for her family's poverty: "We ain't got much, but you're welcome to what we have." And indeed, the family was hard-pressed; Merrick calculated their total income at $26 a month "net cash income," a low number even for a Jamaica farmer. Merrick was a Yale graduate and the son of a business tycoon; he knew what "net" was. But as the anecdote makes clear, he also knew about the critically important non-cash sources of sustenance it did not include. The Christmas table of these tenant farmers was lavishly provisioned, with "roast turkey and roast pork, done to a turn, all you could hold; also five kinds of vegetables from their own supplies, three kinds of pie, two kinds of cake, butternuts, maple sugar, hickory nuts, butter and cream in profuse quantities, and coffee."[21]

A close reading of Merrick's list suggests what he was really writing about: not so much a holiday feast as a survival strategy. Like many of even the poorest Jamaica residents, this family evidently kept a cow ("butter and cream in profuse quantities"); raised poultry and a pig or two ("roast turkey and roast pork"); cultivated a garden and canned and cellared the produce ("vegetables from their own supplies"—see Figure 7); and perhaps tended a few fruit and nut trees (although apples, butternuts, and hazelnuts could also be gleaned from roadsides and field edges). This Christmas dinner would have required cash only for the flour used in the pies and cakes, perhaps a little cane sugar to supplement their own maple sugar, and the coffee that put the finishing touches on the feast.

Even readers far removed from the hill farms of Vermont would have recognized the implicit message in Merrick's list: better to embrace a low "standard of living" in a backroad farmhouse that had food on the table and heat in the kitchen, than to stay in town where your job ended six months ago, and where food, heat, and housing required cash you no longer had. That was the point Professor Peach had tried to make to his colleagues back in 1929 when he spoke about the value of a home from which one "can never be starved or frozen." It was also one of the messages that Jamaica residents had tried to send to the eugenics investigators in 1930. Now it was making sense to more people.

Back to the Farm

When times are bad, back-to-the-land writing can convert readers and encourage chain migrations. A *New York Times* reviewer of *From This Hill Look Down* commented sympathetically on Merrick's project, describing him as one of the "new pioneers" created by the Depression, part of a movement back to the "worn hills of Vermont," to the very farms "which seemed destined a few years ago to revert to wilderness."[22] And for a few years, it did indeed look as if the tide of population was turning back to the "worn hills of Vermont"—in fact to the very "abandoned" hill farms and declining villages that had been the subjects of so much handwringing just a few years earlier. Even on the national level, the reversal was clear: migration out of the cities began almost concurrently with the stock market crash itself.[23] The number of farms in Vermont increased by almost 9 percent between 1930 and 1935. In Windham County, where Jamaica is located, the number increased by 15 percent.[24]

Perhaps the best-known literary figure to go "back to the land" in northern New England was Robert Frost. Frost owned several northern farms over the course of his life, beginning back in 1900 with a ten-year farming venture in Derry, New Hampshire (during another period of back-to-the-land enthusiasm). He farmed in Vermont's Bennington County from 1920 to 1929, but his longest association was with a mountain community to the north of Jamaica. Ripton was one of the best-known examples of a declining hill town, both because its close association with adjoining Middlebury College gave it press attention and also because it had lost an incredible 43 percent of its population between 1910 and 1920 alone. In 1926, Frost helped found the Bread Loaf Writer's Conference in Ripton; he lived there seasonally, still doing some farming, until his death in 1962.

As early as the publication of his collection *North of Boston* in 1914, Frost's poetry was shaped by his intimate familiarity with the northern New England landscape. His first major Depression-era work was "Build Soil: A Political Pastoral." Frost read it first at Columbia in 1932, still a few months away from the election that would bring Franklin Roosevelt to power. "Build Soil" is a text with many layers, written in a mocking form of classical discourse as a dialogue between two characters in Virgil's *Eclogues*. At the time of its publication in 1936, it was understood to be a rather shocking anti-New Deal political manifesto. In large part, the poem is clearly about poetry itself. But its controlling metaphor is agricultural, and it is deeply rooted in Frost's personal experience with hardscrabble farming.

In one passage, the narrator advises a city refugee to return to the land and take up subsistence farming:

> You shall go to your run-out mountain farm,
> Poor cast-away of commerce, and so live
> That none shall ever see you come to market—
> Not for a long long time.

He even offers specific agricultural advice suited to an economic crisis. It turns the expert agricultural instruction of a progressive generation upside down:

> Plant, breed, produce,
> But what you raise or grow, why feed it out,
> Eat it or plow it under where it stands
> To build the soil.

Forget about new markets and technologies, the character counsels his friend. Revert to the older set of farming principles once despised by forward-looking people: Provide first for your own needs. Take the long view. Market the surplus cautiously, if at all: "The moral is, make a late start to market."[25]

The multiple political and artistic messages embedded in "Build Soil" are matters of debate, but the agricultural advice Frost inserted into the poem is not difficult to understand. Indeed, it is not much different from the back-to-the-land advice making the rounds in prose form. Elliott Merrick advised his readers in 1934 to "live on what they can raise, make their homes out of logs or boards or sod or canvas," relying only on their own efforts.[26] The eminent journalist Dorothy Thompson weighed in with similar advice in a 1937 article in her column in the *New York Herald Tribune*. Thompson spent summers in the town of Barnard—one of the "problem" hill towns listed in the 1929 thirteen-town study—together with her husband, novelist Sinclair Lewis. She based her advice on a neighbor's story, much like Merrick's: "When the depression came and he lost his job . . . he came back to the land, with his wife."

In the beginning, Thompson explained, the family had "what looked like less than nothing," but it took only a few years to turn things around, all by following Frost's advice and "making a late start to market." Like Merrick, Thompson inserted a long list of good things the family had acquired: "a team of good horses, a herd of eight or ten cows, fine stands of corn, . . . an orchard of young apple trees, . . . eggs and poultry for his table, . . . and a cellar stocked with food." Most important, Thompson's neighbor had also acquired the kind of self-respect so desperately at stake in those years of unemployment and loss: "In terms of cash income he certainly belongs to the 'underprivileged,' but I would advise you not to suggest that idea to him. He would not like it."[27] In these phrases one can perceive an almost total reversal of attitudes toward the self-sufficient, diversified, small-scale farming associated with the old-fashioned methods of the hill farms.

Just months before Walter Hard's "Vermont as a Way of Life" was published, historian and public intellectual Bernard DeVoto issued his own manifesto: "New England: There She Stands."[28] DeVoto announced a startling discovery: in that year, when "panic possessed America . . . New England wasn't quite so scared. The depression wasn't quite so bad in New England, despair wasn't quite so black." All those decades of decline had evidently made New Englanders strong: "By the granite they have lived on for three centuries, tightening their belts and hanging on." DeVoto wrote "New England," but he meant Vermont. Like Dorothy Thompson, he based his argument on his experiences with a farm next door to his summer home, in Morgan, far up in the sparsely populated northeastern corner of Vermont.

Visiting his hill farm neighbors, DeVoto "came away with a dazed realization" that he had encountered a household that was "wholly secure." The family's income had been small before the crash; it was almost nonexistent now. Yet they lived well (at least DeVoto thought so) on what they could grow and make for themselves, getting by in much the same ways the residents of Jamaica had reported to the eugenics investigators a couple of years earlier: they canned vegetables, hunted game, tapped maples, milked a few cows, and worked at odd jobs. To DeVoto, however, the farmer next door was the very reverse of the "planless," passive hill farmers depicted by the eugenicists. He looked now like "a free man, self-reliant, sure of his world, unfrightened by the future."[29]

Clearly, there was a political element to these ways of interpreting Vermont. To be sure, neither DeVoto nor Thompson could be characterized as a conventional anti-New Deal conservative. (The jury is still out on Robert Frost.[30]) But while neither was a fan of untrammeled big business, both were deeply uncomfortable with the rapidly growing power of the federal government at a time when the rise of fascism in Europe cast a pall over such "big government" programs. DeVoto's politics were idiosyncratic, ranging between "cranky conservatism and maverick radicalism," as the distinguished historian Louis Masur described them.[31] Thompson was particularly worried that the New Deal looked too much like the rising Nazi power she covered as a journalist. Both of them recognized the urgency of the needs the New Deal was designed to address, but they also hoped that an effective response to the Depression might be found outside the reach of centralizing power grabs.[32] Searching for those solutions in the structures of self-reliance rooted in older patterns of community, they found that Vermont's hill farmers and their new back-to-the-land neighbors offered them a perfect model. Their way of thinking was shared by many of the important actors in the Depression-era cultural transformation of Vermont. They would soon encounter a Vermont politician who spoke for them, and who would make his defense of small towns and small farms the foundation of his political career.

The Rise of George Aiken

In the early years of the Depression, most of Vermont's political leaders continued to support the idea of removing hill farmers from their land. On the federal level, one of the earliest agricultural programs of the New Deal was a plan to remove so-called "submarginal" land from production and resettle the farmers elsewhere.[33] (The technical economics term "submarginal" became ubiquitous in New Deal plans, now as a catch-all term for almost any farmland that had poor people living on it.) Despite Vermont's popular reputation as staunchly anti-New Deal, the progressive wing of the dominant Republican Party was generally inclined to support such large-scale government efforts at the state level. So, it seemed all but inevitable that the state authorities would cooperate with the federal plans when they came around. After all, a similar project had already emerged from the homegrown planning efforts generated by the Vermont Commission on Country Life. The New Deal plan simply offered federal funds to pay for relocating the selected farmers. Implementing it would have been the natural culmination of all the state's studies and reports of the 1920s.

In the early 1930s, in fact, some type of removal plan enjoyed the support of virtually the entire progressive establishment of Vermont: two of the state's largest papers, the *Burlington Free Press* and the *Brattleboro Reformer*; the state Grange and Chamber of Commerce; the head of the Agricultural Extension Service at the University of Vermont; and the Commissioners of Agriculture, Forestry, and Fish and Game.[34] The plan's supporters agreed that removing farmers from hill farms would be good for the state in almost every way imaginable: good for the promotion of tourism; for the development of forestry and its attendant industries; good for the establishment of parks and recreational areas; good for state and local budgets; and, in the long run, even good for the farmers.

In 1934, when the federal Agricultural Adjustment Administration delivered its official plan to buy out some of Vermont's "submarginal" farms, the plan was supported by Governor Stanley Wilson, who set the wheels in motion to get state approval. (Wilson was a progressive Republican who supported more active state government, including state control of highways and a state income tax.). But by the time Wilson's successor was inaugurated the very next year, things were suddenly looking quite different. That was not so much because of Vermont's new governor Charles Smith, who initially supported the buyout project, but rather because of the state's new lieutenant governor, who was vehemently opposed to it. The new lieutenant governor was George Aiken. He would go on to become governor and then U.S. senator. But before his election as lieutenant governor, he had been Speaker of the Vermont House of Representatives, and from that position of influence, he had begun to work against the buyout plan.

In time, Aiken's able political maneuvering would convince many former supporters of the project that removing Vermonters from hill farms was an attack on the most cherished values of the state.[35] After the election, the state legislature began to show signs of coming under Aiken's influence. They followed up their preliminary approval of the buyout plan with the usual procedure: appointing a board to consider the details. But something had clearly changed, as became clear to those in the know when it turned out that it was the new Lieutenant Governor Aiken who had been appointed chair of that committee.

Aiken later told his own amusing version of the story, suggesting coyly that there was perhaps a "bit of irony" in his appointment to the committee, because his colleagues knew that he was living on what he termed "a very submarginal farm." There was a little poetic license in that description. Aiken had grown up on a farm in Putney, twenty miles downhill from Jamaica—a town with hills, to be sure, but not exactly a hill town. Aiken was not exactly a hill farmer, either: he operated an orchard and nursery on land that lay between the richer Connecticut River valley farms and the more rugged hill towns to the west. Nonetheless, it was true that he knew the hill farms well. In an essay called "Not So Submarginal," he suggested that the representatives of the New Deal agencies did not.[36]

Aiken's account of his war against the submarginal buyout program combined a frank acceptance of the reality of poverty in the hill towns with a derisive dismissal of federal claims that the poverty was causing what Aiken called a "great degree of unhappiness." (Like Dorothy Canfield Fisher and Bernard DeVoto, Aiken pointed out that hill country Vermonters had already been quite accustomed to poverty before the Depression came along.) He acknowledged that those hill farmers faced real hardships, but he described them with a wry understatement that was becoming his signature style: the land was "rather" stony, the modern conveniences, "few." Moreover, the "boys from Washington" were missing a key part of the story. Hill farmers themselves "protested . . . that they lived on these poor, rocky farms *from choice* [my emphasis]."[37]

Kindly offering to help the federal agents come to a better understanding of why people made such choices, Aiken outlined their reasons. Like so many others, George Aiken associated the hill farms with intangible qualities: "some folks just naturally love the mountains, and like to live up among them where freedom of thought and action is logical and inherent." But unlike many others, Aiken based his rhetoric on an intimate understanding of the landscape that supported those qualities. Hunting and fishing were better in the hills. Potatoes and apples preferred the high ground and clean air. The grass was lusher and more nutritious in the upland pastures. At heart, Aiken's account echoed a much older defense of the essential nature of Vermont hill soils, hearkening back to the nineteenth-century Board of Agriculture speakers' faith in the soil: "The land on these hill farms doesn't wear out with use and don't let anyone tell you it does."

Aiken's description of the subsistence practices of the hill farms was equally well informed. Hill farmers were able to get by with so little cash, he explained, by burning their own wood and by growing their own food, putting up "hundreds of quarts of canned vegetables and fruits." The farmers raised their own animals, he added, even specifying which ones: "at least one cow giving milk, a calf and a pig growing up and a flock of chickens sustained by scraps from the table." For the little cash they needed, they might sell ferns, or Christmas greens, or make baskets (a new specialty in Putney); they might entertain summer visitors or campers, or of course sell maple products. Moreover, most men worked off the farm part-time, Aiken pointed out—without any of the negative judgments of the eugenicists—"in the maintenance of highways," or in a "lumber pulp job." As he described them, those were all simply ways of making a living, not signs of genetic decay or social decline.

Perhaps more important still, Aiken knew his historic moment. He pointed out that all the arguments used by federal authorities about "the desperate condition of the people in the hill towns" were "based on the census figures of 1929." The Depression had changed everything. In his own neighborhood, Aiken had already noticed a sizable "influx of people from the industrial centers to the mountains." Country people "who had been lured to the big cities years previously by high wages and salaries and bright lights" now "felt their thoughts and steps turning back to the farm . . . , back to where there would be a roof over their heads." Not only returning natives, but retirees found it was less expensive to live in the country, and there were "professional people, hundreds of them—writers, teachers, artists, and others, who found that they could do their work as well or better amid congenial rural conditions."

Where would such people want to live, Aiken asked with a rhetorical flourish. Why, "In the small villages and the so-called abandoned farms among the hills."[38] His careful account of the numbers and prices of properties sold to out-of-staters became the backbone of his final argument against federal submarginal buyouts: the hill farms were quickly becoming desirable as second homes or farms for newcomers.[39] The deal on offer from the Resettlement Administration would have committed the federal government to paying to relocate the chosen hill farmers—but in return they would retain ownership of the land given up. It would simply be poor business practice, Aiken argued, to let the federal government take control of such a valuable resource, at just the point when a discerning eye could see it was about to become more profitable.

It did not take Aiken very long to win this battle. In the process, he made a lasting reputation for himself as a champion of the Depression-era version of Vermont values just as they were taking shape. From then on, at least in this matter, Aiken would be the chief standard bearer of the "Vermont way." He went on to become governor and in 1940 was elected to the United States

Senate, where he remained for thirty-four years, becoming in the end a revered statesman whose influence extended decades beyond his active officeholding. This first big victory was a defining one, in part because of how Aiken cast the conflict. Carefully avoiding confrontation with the many Vermonters who had put years into the project of hill farm removal, he instead portrayed the whole idea as the product of federal overreach. Rather than confronting the Vermont Commission on Country Life, the newspapers, the Grange, the agricultural establishment, and his fellow progressive Republicans, he placed the "boys from Washington" in the spotlight. It is a tactic that would become typical of Aiken's approach to more high-stakes controversies in the years to come.

Who is Perfectly Contented Now?

Perhaps George Aiken would have met with more effective resistance if it were not for the harsh realities that undermined the efforts of the proponents of the buyout. During the debate over the federal submarginal buyout plan, the newly created Vermont State Planning Board weighed in on the side of removal. With funding and consultants from the National Resources Planning Board, the committee published a booklet in 1935, designed to convince recalcitrant Vermonters to support the federal submarginal buyout offer. The *Graphic Survey: A First Step in State Planning for Vermont* tried to reassure its imagined prickly Vermont readers, but the writers quickly ran into difficulty. Attempting to use Aiken's style of rhetoric, they praised Vermont's old-fashioned virtues: "Although pioneering days are over, much of the same quality remains in the population of Vermont today." But the booklet followed those flattering comments with an extraordinarily tone-deaf non sequitur: "Therefore, it is quite natural and appropriate that the government should set up an official State Planning Board to look after the interests of the State and to provide a permanent method of fostering the welfare of its population."[40] Those words—"foster," "welfare," and "look after"—cannot have been reassuring to readers with fears of federal overreach.[41]

An even more important problem with the message of the *Graphic Survey* was that it was working against facts that seemed to speak for themselves. Even the "graphic" part of the message—the maps—seemed to present a message the authors did not intend. Two in particular made a powerful if unspoken case against the removal of the hill farms. They depicted the geographical distribution of "cases on relief" in March 1935 (see Figure 8). 10 to 15 percent of the population of the industrial centers of Rutland, Burlington, Barre, Bennington, and Springfield were receiving some type of federal "relief."[42] In the smaller mining and stone-cutting towns—Poultney, Hardwick, Pownal, Proctor—the situation was worse. There, over 15 percent of the population were on relief.

In stark contrast were a handful of towns that reported no cases on relief at

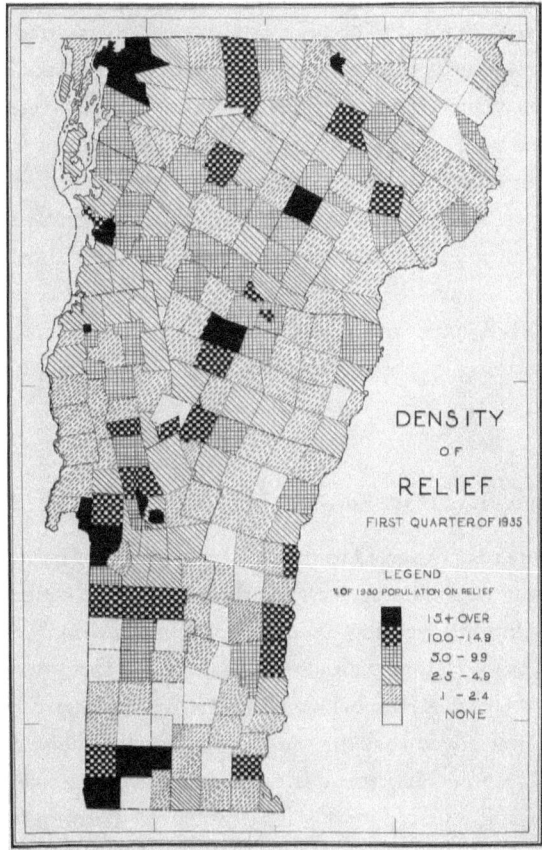

FIGURE 8. "Density of Relief." This map was created by the Vermont State Planning Board to show the percentage of people receiving government "relief," or aid, in each Vermont town in the first quarter of 1935. Looked at another way, it suggests that the most isolated and mountainous parts of the state were experiencing the least hardship. Jamaica appears near the center of the cluster of towns in the southern part of the state, colored white to indicate that no people in town relied on government aid. Source: *Vermont State Planning Board, Graphic Survey, A First Step in State Planning for Vermont*, a report submitted to the Vermont State Planning Board and National Resources Board, 1935, 41. Silver Special Collections Library, University of Vermont.

all, standing out as unsullied white spots on the maps. The largest cluster of those towns was in the southern Green Mountains—the geographic center of all the worst hill town problems of "abandoned" farms, "submarginal land," and "inefficient" farmers—including Jamaica and most of its surrounding towns. Jamaica, Wardsboro, Windham, and Stratton reported no cases of government aid. Winhall, Townshend and Londonderry—and Governor Aiken's hometown of Putney—reported somewhere between 0.1 and 2.4 percent of their populations on relief. A second conspicuous white area was a handful of very small towns directly adjacent to Morgan in the sparsely populated northeast corner of the state, where Bernard DeVoto had encountered his self-reliant neighbor Jason.[43]

Of course, there are many possible explanations for why the residents of these small towns were not receiving federal aid. Possibly their local networks took care of their own; perhaps shame kept them from asking for help. Rural Vermonters may not have been ashamed to be poor, as Walter Hard asserted, but that did not mean they were not ashamed to be "on relief." George Aiken himself wrote a scathing essay about requirements for a jobs program that were so insulting and invasive that only two men in his town accepted the work, although many desperately needed it.[44] But whatever the reasons might have been, the maps carried one message that could not have been missed by anyone who picked up the *Graphic Survey*. No matter how marginal the land, how monotonous the diet, how inadequate the clothing, how few and far between the luxuries, the people of the hills were not—like so many others in those dark days—homeless, cold, or standing in a bread line. It was a point that would have been difficult to miss in 1935.

Some Experts, Too, Have Second Thoughts

Most agricultural experts were not converted to the new enthusiasm for backwoods subsistence farming. But there were signs even in those specialized circles that the Depression was forcing at least a temporary change, in tone if not in actual strategy. Some agriculture economists took a second look at their own certainties about the future of farming. Everywhere in northern New England, agricultural economists acknowledged that the most commercially oriented farmers felt the impact of the Depression first and felt it most deeply. More than the other states, Vermont was dominated by the dairy industry.[45] Before the Depression, the state's agricultural experts had almost unanimously supported a shift to commercial production of milk. Then, between 1929 and 1932, milk payments to farmers dropped by half. The long-term pattern was even worse. In 1939, when most of the economy seemed to be recovering, the milk price had still risen by only 25 percent.[46]

Some experts wavered, painfully aware that the farmers who had most enthusiastically followed their advice to specialize, commercialize, and centralize were now paying a price for that choice. In his report on a study of the impact of the 1930 price slump, University of Vermont agricultural economist Sheldon

Williams spelled out what it had meant for his clients. Had it not been for their insistence on continuing their outdated farming practices, he acknowledged, there would have been much greater suffering on those farms. "In all areas," he found, "the farms having the smallest percentage of receipts from cattle"—that is, the most diversified—did better than those "deriving the greatest portion of their income" from dairying. Prices for poultry, wood, and maple also declined, but not as dramatically as milk prices, so "the advantage of diversification was greater than usual."[47] The farmers who weathered those first few years with the least difficulty were those who had somehow managed to ignore all the advice from the state.

Reporting on a 1932 study, Williams and his research partner did their best to maintain the orthodox position that bigger dairy farms were better, even though the bigger farms were harder hit by the Depression. The situation was temporary, they reminded readers. And they advised farmers to continue to increase the size of their farms, provided that they "recognize clearly that the large farm business is particularly vulnerable in times of general agricultural depression." They warned farmers to "take every possible precaution against getting caught out on a limb." But that advice must have rung a little hollow in 1936, when the report was published: they offered no advice for how to avoid getting "caught"—probably because they did not have any.[48]

To be sure, things were not rosy for old-fashioned farms, either. Windham County's farmers were far more subsistence-oriented than those in the state's commercial dairy regions, and thus probably less susceptible at least to the initial impact of the Depression.[49] But (as the pre-Depression investigators had discovered) most of the general and "self-sufficing" farmers in Jamaica and other similar places depended on the patchwork of jobs in the woods and sawmills for cash. Those jobs were increasingly scarce in the 1930s.[50] At the same time, these kinds of diversified farms often sold their surplus eggs, butter, wood, or potatoes in local markets, and those were also facing declining prices and demand, as their neighbors tightened their own belts and made do with less. Very few farm households could simply do without all market transactions and rely on their own supplies, and there must have been relatively few families feeling as secure and independent as some observers imagined them to be. Even so, the perception that old-fashioned farms were more resilient than modern farms seemed to make at least some northern New England agricultural economists a little less certain of the way forward.

Placing a New Value on Security

One major study surveyed an area of declining population and "abandoned" farms in central New Hampshire. The study was modeled on the thirteen-town project, and it followed the methods employed by Lemuel Peet's team. In 1934, the two foresters who conducted the study reported on what they had found, and proposed that the state should move the remaining population off the farms and take ownership of the land. But their use of a term coined by New Deal government workers hinted at a partial shift in perspective: the farmers on these hill farms were now described as "stranded." That term carried a lot of meaning at the time: it linked New Hampshire farmers to coalminers in the Appalachians, or steel workers in Pennsylvania, where entire towns full of workers were "stranded" because of the failure of one company. Most of all, the term implied that they were in trouble through no fault of their own. While earlier investigators had scrutinized the behavior and character of hill farmers to assign responsibility for their problems, these scientists put the blame squarely on outward circumstances.

Harry Woodworth, the agricultural economist who led this New Hampshire study, injected a note of deep uncertainty into the study, questioning the entire premise of such efforts—efforts he had been involved in for years. After describing the soil types and agricultural history of the area, and the incomes and productivity of the residents, Woodworth ended with a startling question: "Who is to judge," he asked, "whether this situation or that is best for an individual family?" Perhaps for some people, life on the submarginal hill farms was better: "Must all conform" to modern standards with their "artificial" and "complex" requirements? Or was it possible that for some, "nonconformity to the general standard brings a richer content of living?"[51] Before the Depression, the question had never come up in quite that way.

In the end, Woodworth made a powerful argument against his own original proposal to remove some farmers from their "submarginal" land. Agricultural economists, he noted, calculated productivity in terms of what a healthy man could do at the height of his strength: one "who can aggressively push productive processes." That left out of consideration older people and those who "lack the physical or mental ability to operate a large commercial farm." Those small subsistence farms might be the best that some people could manage. At the same time, Woodworth's reconsideration was influenced by his discovery that the remote "submarginal" farm areas the state was hoping to clear were no longer losing population but were now attracting a wave of new Depression-era settlers. Some people, he now realized, had few alternatives, and none that guaranteed basic food and shelter. "Security is a great human goal, and after all is part of the content of living. With many, especially older people, it is often the dominant factor."[52]

A study in upstate New York showed similar signs of having been changed by the impact of the Depression. This one was designed to evaluate the state's already existing program to buy out farmers on submarginal land. Overall, the investigators found that the state buyout programs had been successful, but their data also turned up a new problem: 49 percent of those living on the most marginal farms had been there fewer than ten years, and half had come from non-farming jobs. The researchers discovered that, in this area, too, it was likely that desperate, jobless people would be moving to farms in this region faster than the state could buy out those who were already there.

In recognition of that reality, the investigators included some words addressed to those who wanted to stay on their submarginal farms despite expert advice to the contrary. If the farmer was young and strong (and had some training as a farmer), they explained, their advice would still be to sell out to the state, save their money, and buy a farm on better land. But for the people who wanted to stay on those marginal farms, their advice was much more old-fashioned. In fact, it might have been written by a Vermont Board of Agriculture member back in the 1870s: be very careful about borrowing; hire as little labor as possible; raise chickens. Finally, in the last paragraph of their 24-page report, the authors concluded: "The individual who is farming in land classes I and II [in this system the worst soils] should obtain as much of his own living from the farm as possible." And the specifics would have been familiar to anyone who had read or heard anything about self-sufficing northern farms: "He should have a good garden and store and can enough vegetables and small fruits so that these will be available at all times of the year, and he should "produce enough milk, butter, eggs and meat for his family."[53]

"In the Present Emergency"

There was one group of experts who had been sending exactly that message of household self-sufficiency even before the Depression. These were the home economists at the University of Vermont's Agricultural Experiment Station.

(Their work is discussed in chapter 5). All of them were women, and they seem to have operated a little apart from their male colleagues, making up what amounted to their own scholarly circle. In the context of their research, household self-sufficiency had always seemed like a good idea. The Depression only intensified their interest in safety first strategies on farms. In 1932, for example, Marianne Muse commented that "self-provisioning" was both practical and comforting: "Certainly these home-grown products enable many farm families to live reasonably well even in times of depression or crop failure." But they also provided a psychological benefit: "a sense of security sadly lacked by city industrial workers on the same income level."[54]

By 1935, the home economists were fully committed to responding to the needs generated by the Depression. That year, Muse supervised a master's thesis that investigated the question of exactly which goods should be home-produced rather than store-bought. Dorothy Emery's thesis began with the comment that the experts were now in agreement that all farm households should produce most of their own food, at least for now—"in the present emergency."[55] Emery provided an extensive list of the types and amounts of foods that could reasonably be produced on the farm.[56] It went almost without saying that Vermont farmers should produce all their own milk. (Emery recommended budgeting a generous five cups per day per person.) But she also suggested that farm families provide as much beef, pork, poultry, and eggs as possible for their own tables. And she generated a daunting list of thirty-five vegetables that could be raised in Vermont, several of which—such as "kohl-rabi" and "egg plant"—might have sounded a little exotic to many farmers. Emery recommended that farm families provide all their own fruit, including ten varieties of cultivated fruit and three wild ones.[57] Butternuts were available everywhere in Vermont for the picking, she added, and could make an important part of the diet, as could wild greens in springtime. Emery's list was assembled in the form of a very long chart.[58] But in its own way, it bears a resemblance to Eliot Merrick's effusive depiction of his abundant home-made Christmas dinner with the tenant farmers. It was a list that carried a reassuring message to its readers: there would be plenty to eat on the farm.

By the late 1930s, one might travel from Jamaica in any direction and run into someone who was making it their business to articulate and defend the advantages and virtues of the hill farms and their ways of life. Indeed, a map of their whereabouts would make Jamaica look almost like the epicenter of that new cultural wave. Walter Hard lived in Manchester, just over the mountains to the west. Dorothy Canfield Fisher lived a little farther southwest in Arlington, and the novelist Zephine Humphrey and her husband, landscape painter Wallace Fahnestock, lived in Dorset, the next town. To the north of Jamaica was Landgrove, where Samuel Ogden had bought up and reorganized an entire town and was now refurbishing farmhouses and barns for an assortment of classical violinists, architects, and writers who wanted a taste of the hill farm life.[59] Just east of Landgrove was Weston, where Vrest Orton founded a series of enterprises designed to resuscitate small-scale industries that he hoped would bring people and jobs back to the hill towns. (One of them became the Vermont Country Store, still in business today.) Due east was Westminster, the home of Charles Morrow Wilson, who wrote about his back-to-the-land experiences in *Country Living Plus and Minus*. And to the southeast was Putney, the home of George Aiken. On a national scale, the Southern Agrarians at Vanderbilt University garnered more attention with their 1930 collection *I'll Take My Stand*, articulating a similar region-based defense of old rural ways. But in the north, it would be these interlocking circles of Vermont writers who took their "stands" for small, diversified, un-industrialized farms.

Return to Contentment

Of all the new interpreters of Vermont, the one with the deepest roots in Jamaica was Muriel Grout Follett, whose description of deer hunting was discussed in chapter 6. Born and raised on a farm in East Jamaica, Follett married and moved to a farm across the border in the next town over, Townshend. She was an aspiring writer when she met John and Marion Rice Hooper, the editors of a new regional press based in Brattleboro. The Stephen Daye Press, originally founded by Vrest Orton in 1931, was a home for the new regionalist literature emerging in Vermont during the Depression. (It published, for example, Elliott Merrick's 1934 *From This Hill Look Down*, Charles Morrow Wilson's 1938 *Country Living Plus and Minus*, and George Aiken's 1936 *Pioneering with Fruits and Berries*, a product of his first occupation as a nurseryman.[60]) Follett had been hoping to write a novel for the press, but the Hoopers convinced her instead to publish her diary, probably with an eye to the Depression-generated enthusiasm for old ways of farming.

The diary became *New England Year: A Journal of Vermont Farm Life* (1939). At first glance, *New England Year* seems to be just another "country life" book, with the typical amusing anecdotes and ruminations on the timeless beauties of rural life. But those kinds of books generally recounted the adventures of well-to-do exurbanites, as did Frederick Van de Water's *A Home in the Country* (1937), which described his move from Manhattan to Dummerston (two towns east of Townshend). Muriel Follett's book was something very different, in one way, at least: it was the work of a woman born and raised—and still living on—a farm in a hill town. Written only a handful of years after the eugenicists visited, it is useful to imagine Follett's book as one last message from Jamaica. This time, rather than simply responding to the questions of the experts, Follett was able to frame that message in her own terms.

In some ways, perhaps, Muriel Follett was ill suited to speak for the hill farms. The Folletts' place was very large—some 550 acres—and they were clearly prosperous. (Perhaps as important, her family occupied a respected position in Jamaica, and both she and her husband held college degrees.). The farmhouse had electricity and labor-saving devices still uncommon in hill farm communities. Mrs. Follett had "a girl" to help her around the house, and Mr. Follett kept one permanent hired man and hired others for big projects. In many ways, too, theirs was a thoroughly modern farming operation. Mr. Follett milked fourteen cows (not a large herd by commercial standards, but larger than most in Jamaica and Townshend); and he was one of the seven directors of the Cooperative Milk Plant in Brattleboro (a growing concern that was consolidating the smaller cooperative plants in nearby small towns).

The Folletts also demonstrated a clear understanding of how to make the most of the emerging "brand" of Vermont. The family produced a range of charmingly packaged maple goods from syrup to "soft pail sugar" and hard candies. Mr. Follett drove his carloads of maple products all the way down to Connecticut to sell direct to his customers, because, as Mrs. Follett recorded, "his dry Vermont twang" was good business: it offered customers the "best possible proof that they are getting a genuine Vermont product."[61]

Despite its size and sophistication, though, the Follett place still in some ways resembled the Jamaica farms described by the eugenics investigators in 1930. More than anything else, the sheer diversity of the family's efforts pointed to an older style of farming. Like their neighbors on more modest farms, the Folletts tended to a multitude of disparate tasks. The farm's chief crops included a traditional mix of maple, wood, and milk. The family's subsistence efforts took up much of their time. Mr. Follett cut his own wood to heat the house in winter, wood to fire the evaporator for maple production, and logs to be taken to the sawmill and returned to the farm for building projects. They still killed their own pig each year, slaughtering it, butchering it, and processing it: Mrs. Follett and her "help" made sausage, bacon, ham, and lard. Like her less prosperous neighbors, Mrs. Follett did prodigious amounts of food processing for winter, canning peas, corn, broccoli, blackberries, applesauce, pickles, venison, chard, beet greens, and currant juice: forty-seven jars in just two days in July, followed by twenty-five more in late July and forty-eight in September.[62]

On top of the main crops of milk and maple, and in addition to their provisions for the household, Mr. Follett worked off the farm, drawing lumber for other farmers with his team of horses. And in the spring of 1938 the couple set out four hundred strawberry plants and thirty-seven rows of shelling peas to sell in town (to summer residents, mostly, since "most of the natives have their own gardens."[63]) All this diversification, moreover, was not just the product of some vestigial tradition, but a carefully reasoned decision. Recounting a Home Demonstration program discussing migration from farms to town, Mrs. Follett reported that the audience had agreed on the principle: "We decided that diversified farming in this section of the country is about the only way to balance the budget"—a point that might have been instructive for the investigators to hear a few years earlier. The truth was that "even with an income from diversified farming, the present day farmer is a long way from being well off."[64]

Elin Anderson and her coworkers had worried that Jamaicans' diversified work patterns allowed them too much control over their time—a freedom the eugenicists believed encouraged laziness and passivity. Follett agreed that farmers possessed more control over their working lives than most did, but for her that was a matter for celebration, not worry. While "chores tie us down morning and night," she wrote, "during the day we are free to decide for ourselves what we will do—more or less."[65] Workplace autonomy was a natural and important advantage of farming. In an echo of Robert Frost's poem "A Time to Talk," she pointed out how that advantage operated: "in what other business can one stop in the middle of the day ... to talk with friends?"

Mrs. Follett's language was as demure and inoffensive as it could be, but listening closely, one may still hear in her words an implicit challenge to the way of thinking that had brought Elin Anderson to her town. Too much contentment had been the centerpiece of Anderson's diagnosis of Jamaica's failures. But Muriel Follett did not seem to think there could be "too much" contentment. She began her central message on a bit of a sentimental note, the kind of banal greeting card message one might expect in a standard "country life" book: "It isn't what you have but how you feel about what you have that makes for happiness."

(My own copy of the book has that passage inscribed on the flyleaf by a former owner.) As she continued, though, her point became a little sharper: "Farm living brings big returns in everything but money." Finally, almost as if she were answering Elin Anderson directly, she drew her conclusion: "Maybe it is because farmers as a group are too content with their lives to fight for money, that they have no more." In Anderson's analysis, that had been what was wrong with the hill farms of Jamaica. To Muriel Follet, that was the very thing that was most right. She concluded, with characteristic modesty, "It's an interesting conjecture, if nothing else."[66]

EPILOGUE

Returning to Jamaica

In the end, no farm removal plans materialized (in Vermont, that is).[1] But the Depression-generated surge of support for old ways of farming did not last, and it did not prevent longer-term forces from continuing to operate in the hill towns. World War II's sudden full employment, booming food markets, and more active government aid and regulation, would immerse rural people in the modern world, both as producers and as consumers. The war would pull rural people from farms in greater numbers than ever, into military positions and other jobs across the nation. And in the years after the war, agricultural experts would once again wholeheartedly embrace and promote the continued industrialization of farming.

For northern farmers, the postwar years would be decisive, conclusively separating farm households who were able and willing to modernize from those who would not or could not. In upcountry Vermont, as in many other places, old ways of farming were finally abandoned or forced out. The modernizing plan for American farms—more expensive and complicated technology, more use of fossil fuels, more specialization—would overpower nearly all lingering resistance. By midcentury, it would seem clear that the long struggle between traditional and modern farming was over, and that modernity had won.

The End of Farming

In Jamaica, as everywhere in the hills, the number of farms continued to dwindle. By 1940, the census reported sixty-eight farms remaining, and by 1950, just fifty-four—barely over half the number the census had reported in 1930 (itself a decline by half since 1880). In 1950, the census bureau recognized the changes that were overtaking many parts of rural America, distinguishing for the first time between people *living* on farms and people *farming*. In Jamaica, only thirty of the fifty-four households who lived on farms now reported farming for a living: sixteen were listed as farmers, ten were "sugaring," and four still called themselves "cattle dealers" (a long tradition indeed). The rest were making a living some other way: cutting logs, driving trucks, teaching school. By that time, Jamaica was once again a landscape of wood, stone, and stream. Only 13 percent of its remaining farmland was now in crops.[2] On many farms, fields and pastures had long ago grown back to brush; some former farmlands were now supporting fully matured stands of evergreens and hardwoods.

Yet there was still some continuity. Even back at the beginning of the twentieth century, many Jamaica farmers had spent as much time in the woods as they did in the fields. Over time, more and more households gave up selling potatoes or milk in favor of selling maple syrup or lumber—a continuation of a decades-long process. Most men in Jamaica continued to work in familiar settings: on the farm, on the roads, or in the woods.[3] Households could still make the choice to hold onto the land. They could still rely on extensive family and neighborhood networks. They could even pursue old strategies of diversification. Though their land had become mainly woodlot and sugar bush, they could still grow gardens, chickens, and cows. (In 1945, 58 percent of Jamaica farms still milked cows.[4])

The End of "Decline": Jamaica Rebounds

At the same time, while the numbers of farms were dwindling, the population of Jamaica was no longer declining. In fact, the population had already been stabilizing when the Eugenics Survey came to town in 1930 to study Jamaica's decline. During the 1920s and 1930s, the population remained almost completely unchanged (566 in 1920, 570 in 1930, 567 in 1940). But in the 1940s, it suddenly jumped by 5 percent—the first increase of any size for well over a century. And that was just the beginning. Like many other Vermont hill towns, Jamaica would eventually grow back to nearly double its lowest population. After one more drop in the 1950s, the population jumped 19 percent in the 1960s, followed by a 15 percent increase in the 1970s, 10 percent in the 1980s, and an enormous 25 percent increase in the 1990s. By the end of the twentieth century, the town population would hover around 1000—larger than it had been in over a century.

It was not easy to see at the time, but this change, too, had its roots in patterns that emerged when the population was at its lowest. Already in 1930, the eugenics investigators had observed a significant influx of newcomers in town. A third of the heads of household they interviewed had been born outside Jamaica and its surrounding towns, often coming from other parts of Vermont and northern New England, but also from as far away as Arkansas and New York City. In the Depression decade, many more newcomers replenished the population of the town. The 1940 census reported that 32 percent of Jamaica's heads of household had arrived there over the previous decade.[5]

As in the past, many of these new arrivals were from nearby towns, moving in and out in search of jobs. But just as the eugenics study had found ten years earlier, some of the 1940 newcomers had different stories. As in 1930, some arrived originally as summer or winter residents. Some were seeking an inexpensive place to retire, or a temporary or even permanent escape from city life. Some were those "able but maladjusted people" Elin Anderson had identified, people who "chafed against their fate" in the cities. Back in 1930, such new migrants had reported that they came to Jamaica because they preferred country to city life, because they wanted to farm, or because their health required outdoor work or a mountain climate. There was no second eugenics survey in 1940 to ask new residents why they had come to town, but the pattern of migration itself suggests that some of the new people harbored similar motives.

Because New Deal policymakers were interested in keeping track of the back-to-the-land movement of the population during the Depression, the 1940 census asked residents to report where they had been living in 1935 and whether they had been living on a farm. That information can substitute for the eugenics survey in a small way, because it reveals exactly where the newcomers had come from most recently. Fifty-one new migrant households moved into Jamaica between 1935 and 1940, constituting more than a quarter of the town's total number.[6] Of those new residents, a third came from nearby areas of rural Windham County or rural Bennington County; a quarter came from other parts of Vermont and the rest of northern New England. But over 40 percent came from more urbanized areas to the south: southern New England, New Jersey, New York, and Pennsylvania.[7] That was the statistical bump New Deal policymakers were looking for: a discernible "back-to-the-land" contingent.

In the old neighborhood of South Hill, most people living there in 1940 still had family connections that went back generations. The few people who did move in had long-standing ties to the neighborhood. The Bauer brothers returned home from Staten Island at some point between 1935 and 1940.[8] Another man came from Pennsylvania, but his father had been born in the neighborhood, so he was not really a newcomer, either. Elsewhere in town, however, different kinds of newcomers made their appearance. Some appear to have been retiring from their city occupations. For example, Mabel Anderson and her brother Frederick were both in their sixties when they moved to Jamaica. Mabel had been living in Queens and Frederick in Boston in 1935.[9] After Frederick's wife's death in 1937, he joined his sister on a farm in Jamaica. Down the road was another family with a similar story. Thaddeus Seaman and his family had moved from Long Island, where he had worked as a real estate appraiser for a mortgage company.[10] Sometime between 1935 and 1940, when Seaman was approaching sixty, he bought a Jamaica farm and retired there.

There is no direct evidence to show how either of these families chose Jamaica, but the census contains one hint that they may have been seeking a more rural life. Mr. Seaman's path to Jamaica seems to have been part of his retirement plan, for example, but in 1940, he told the census taker his occupation was "farmer." (His stepson was labeled "helper on a farm.") Mabel Anderson, too, gave her occupation as "farmer" (her brother gave no occupation).[11] And among the newcomers in 1940 were Massachusetts engineer Oliver William Nash and New Jersey salesman Frederick Bartlett. In 1940, they both told the census taker they were now farmers. The pattern extended even to the cases of David and Margaret Newton, who moved to Vermont from Connecticut in 1937. Graduates of Princeton and Vassar, the two came to town to open a school for boys that would make its rural and rugged location a key feature of the education it offered. In the 1940 census, David Newton, too, listed himself as a "farmer," while Margaret Newton was labeled "proprietor of a boys' school."

The decision to take on the title of "farmer" may or may not reveal much about what occupied these new migrants from day to day, but it does suggest that they were part of the widespread back-to-the-land activity the census-takers were attempting to count in the 1930s. The Newton School, indeed, was a conscious part of that movement. The school's rustic environment and remote location were key parts of its appeal to both teachers and students (and the school did produce its own food, so the label "farmer" was not unreasonable).[12] Collectively, these farming newcomers and their children would have a significant impact on Jamaica.

The Radical Farmers of Pikes Falls

Jamaica's most famous Depression-era city-to-farm migrants were Helen Knothe and Scott Nearing. The couple would later come to occupy an almost legendary position in the back-to-the-land movement of the 1970s, but Scott Nearing was more infamous than famous when the pair moved to Vermont. Back in 1915, Nearing's promising career in economics at the University of Pennsylvania's Wharton School had come to an abrupt end when he was dismissed for his outspoken support of protective child labor laws. In 1917, he joined the Socialist Party and began work for its Rand School of Social Science. In 1918, his anti-war pamphlet *The Great Madness* resulted in a highly publicized trial under the 1917 Espionage Act. In court, Nearing presented a powerful defense of the right to free speech and was acquitted, but he was increasingly unemployable. In the 1920s, Nearing left the Socialists, joined the Communist Party, and started writing for the *Daily Worker*, but within a few years he broke with that party, too. By 1930, now middle-aged, Nearing found few opportunities to make a living by writing or public speaking. Helen Knothe had her own story. Raised as a member of the spiritualist Theosophical Society and trained for a career as a violinist, she met the older Nearing a few years after a difficult break-up with the charismatic young leader of the Theosophical movement Krishnamurti. As the Depression deepened, the couple determined to move to Vermont.[13]

The Nearings gradually acquired over a thousand acres in the Green Mountains. They developed a sizable maple syrup business, working the sugar bush on shares with their neighbors.[14] Under Helen's guidance, they crafted a successful marketing strategy, packaging the syrup and the maple sugar in distinctive boxes and bottles and advertising the products to a variety of discerning clients. In the 1940 census, both Helen and Scott were listed as farmers. In 1950, Scott was still listed as a farmer, but Helen, like several of their neighbors, now identified herself as a maple sugar maker.

As it happens, the Nearings do not actually appear on the census rolls of Jamaica, because their land was located just over the border, in the tiny mountain town of Winhall. By 1930, Winhall had almost disappeared into the forest, its population of 212 supporting itself mostly by logging. (Its location in the valley between Bromley Mountain and Stratton Mountain did not yet mean much.) But Winhall, like Jamaica proper, was attracting its own new migrants. As in Jamaica, most were local folks making short moves back and forth across town borders. But in the 1940s, a different set of newcomers was also beginning to appear, drawn in part by the presence of the Nearings themselves. Like the Nearings, these new migrants would embrace the agricultural heritage of their adopted community—in their own way, at least.

To take one example: in 1950, the census reported a new resident on the Nearing property, a man named Richard Gregg, who happened to be a leading figure in the American pacifist movement. In 1934 Gregg had published *The Power of Nonviolence*, the book that introduced Americans to the theory and practice of Gandhian nonviolence. His book would play an important role in Martin Luther King, Jr.'s adoption of nonviolent strategies during the 1955 Montgomery bus boycott. Gregg's 1936 pamphlet *The Value of Voluntary Simplicity* was a similarly groundbreaking introduction of that term and concept into American use. In Winhall, though, the 65-year-old Gregg was not recorded on the census as a writer, teacher, or philosopher. He did not even claim to be a farmer; he told the census taker he was just a "farm hand" on the Nearing farm.

In these years, scores of people with similar convictions were attracted to Jamaica's wild western edge. During the 1940s, no fewer than three small intentional communities were established in the area, loose affiliations of anarchists, Quakers, pacifists, and distributists, each forming a cluster of hand-built homesteads.[15] During World War II and after, draft resisters, too, found refuge there, some after serving prison sentences.[16] A 1948 article in the *Brattleboro Reformer* described the new community: "They Get Away From It All in Pikes Falls." The subheading announced, "Neighbors Think They're Queer."[17]

That article did not make much of the new arrivals' radical politics or of their resistance to the draft, did not name anyone a Communist or draft dodger.[18] Instead, it seemed to find most "queer" the desire of the newcomers for a simple life, "far away from the luxuries and comforts of city life." It appears the reporter received some pushback from readers, because in the next week the paper ran a short piece that intensified its critique: living "a simple life on the land," the writer now made certain to point out, "is utterly impractical." (Anybody who seriously wanted to live off the land would move to a tropical climate anyway.) Eugenicist Elin Anderson would have diagnosed the Pikes Falls settlers as "able but maladjusted." This writer called them "fugitives from reality."[19]

For these new migrants, however, simple living was not in itself the whole point of their experiment in Pikes Falls: it was rather a necessity born of their political choices. In a very real way, their radicalism did make them "fugitives." The idea of "voluntary simplicity" was deeply rooted in Gandhian principles of nonviolence, but of course, it could also be a way of avoiding the attention of the federal government. Several of the residents had been fired from teaching positions and one was even fired from a job at the weather station atop Mount Washington. Since they could not keep jobs out in the "real world," they chose to live in a remote village, building their own shelter and growing their own food.[20] So it was that when Alfred and Norma Jacob bought land and built a house in the area, they put out the word that war resisters were welcome to join them. Thus, after Norman and Winifred Williams came to live there in 1943, and Marshall Smith bought his land nearby in 1946, all three men listed themselves as "farmer" on the 1950 census.

Over time, these migrants would play a major role in Jamaica's re-growth, attracting supporters and curious observers even after the Nearings left for Maine in 1952. Back in 1930, Professor Peach had associated the hill towns with personal freedom, describing an "independence of spirit . . . some sense of freedom that far hill vistas give." In 1938, Governor Aiken had asserted that "freedom of thought and action is logical and inherent" in the mountains. Now, as Jamaica became known to the FBI as the home of anarchists, socialists, and anti-war activists, the old trope of "mountain freedom" took on a new layer of meaning.

As Goes Jamaica, So Goes Vermont: The Ski Industries

In short, Jamaica was becoming what Vermont would soon become—only a little ahead of the rest of the state. The area around Jamaica was among the first in the state to feel the full impact of the skiing boom on the landscape and economy. The ski industry arrived in Vermont in the 1930s, growing out of a larger state development vision with roots in early twentieth-century efforts to promote tourism, hiking, and camping in the mountains. The first ski resort near Jamaica appeared in 1936, at Bromley Mountain in the town of Peru. The first ski trail there was built with the aid of the federal Works Progress Administration, around the same time that similar ski areas were developed in Stowe, Woodstock, and Killington. After World War II, big resorts multiplied. Two towns south, Dover's Mount Pisgah was renamed Mount Snow and the resort opened in 1954. The town of Stratton, directly to Jamaica's west, was already a well-known wilderness destination, as the Appalachian Trail coincided with Vermont's Long Trail on Stratton Mountain's peak. In 1961 Stratton Mountain became a major ski resort as well; the corporation that owned it would eventually expand to take over Bromley Mountain in Peru (two towns north), Magic Mountain in Londonderry (adjoining Jamaica on the north side), and Timber Ridge in Windham (adjoining Jamaica on the northeast).

All the tourism enthusiasts of the 1920s and 1930s were right, of course: tourism, vacation, hiking, and especially skiing did bring business to Vermont. But equally clearly, skiing brought its own problems. The ups and downs of that volatile industry have troubled Vermont's state budget-making for nearly a century now. Real estate booms and busts, sprawl and shoddy development, struggles over natural resources—these problems and opportunities came first to towns like Jamaica, to replace the old problems of the hill farms that had once worried state leaders.

As Goes Jamaica: More Radical Farmers

Jamaica and its neighboring towns anticipated the future of Vermont in another way, too. After midcentury, Vermont's population rebounded dramatically, increasing by 14 percent in the 1960s, followed by another 15 percent in the 1970s. Not all that population boom was caused by "hippie" immigrants, of course, but a significant portion of the growth was made up of young out-of-state migrants. In cyclical fashion, as those new migrants discovered Vermont, the whole state become known as a haven for political radicals, draft resisters, and a new wave of back-to-the-landers.[21] And, as in Jamaica in the 1930s and 1940s, those new migrants often aspired to be farmers.

The connection between radicalism and farming was a close one, both in 1940s Jamaica and in 1970s Vermont. Along with their other countercultural practices, the Pikes Falls radicals brought with them to Jamaica an interest in organic gardening and farming methods, then newly emergent in the United States.[22] (That is one reason why the philosopher and writer Richard Gregg was in Jamaica, working as the "farm hand" of Helen and Scott Nearing, developing new techniques for organic growers.) In the 1940s and 50s, organic growing methods were often viewed as a kind of cult, embraced by people on the radical fringes of American culture. (And indeed, organic advocates did often portray it as a more "nonviolent" kind of gardening and farming.[23])

By the time the new generation of migrants arrived in Vermont in the 1970s, however, organic gardening and farming had become a fundamental tenet of countercultural thinking: not yet mainstream, but increasingly familiar to a broader audience. New migrants to Vermont embraced the opportunity to produce healthier and more natural food for themselves and for local markets, establishing farmers' markets, community supported agriculture programs, and food cooperatives. Growers in the southern Vermont town of Putney (George Aiken's hometown) founded the Northeast Organic Farming Association in 1971; that town's food coop had already been in business for thirty years by then.

Perhaps not many of those newcomers were quite "farmers" in the sense that most people imagined the term in those years. In midcentury, the red barns and black-and-white cows of Vermont's successful commercial dairy farms still dominated the landscape and the imagination of visitors. But the newcomers often put down roots on old hill farms, inexpensive land now partly reforested and little cared for. Their agricultural practices would bear more than a passing resemblance to those that had been standard in hill towns in years past—sometimes, indeed, because the newcomers learned to farm from older hill farm neighbors.[24] Over time, they became their own kind of hill farmers: modest, savvy, and attuned to their land. That convergence would play an important role in the long process of agricultural change in Vermont.

As Goes Jamaica: The End of Farming

The most ironic of historical turns was one that few of the advocates of removal of the hill farms could have anticipated in 1930. The process of "allowing" the hill farms to return to forests, it turns out, would not end with those small backroad farms. For a few decades, to be sure, Vermont was a place defined by its successful dairy farms. Cattle grazed in green valleys framed by now forested hills, creating a recognizable look for a generation of tourists and summer camp kids. Artists who moved to Vermont in the late twentieth century created distinctive styles based on that landscape: the work of Sabra Field, Mary Azarian, and Woody Jackson would be unrecognizable without the barns, silos, valley corn fields, and black-and-white Holsteins that became something approaching a state trademark.[25]

Today, that valley landscape, too, is all but gone in much of the state. By the late twentieth century, ruins of once-prosperous downhill farms were joining those of the "submarginal" uphill farms, now almost buried in the trees. At the same time, more and more of those valley farms were being replaced, not by stands of woods, but by shopping malls and housing developments. By 1993, the National Trust for Historic Preservation recognized the extent of the change by placing the entire state on its "most endangered historic places" list.[26]

A hundred years ago, agricultural modernizers had high aspirations to save Vermont farming from losing out to national and later global industrializing forces. In the end, it seems, they were able to provide no more than a short respite. In 1930 there were some 35,000 farms in Vermont. In 2020 under 7000 remained, and the number continues to decline yearly. For decades, market forces and agricultural specialists alike have pushed Vermont farmers to get big or get out—and for decades that has mostly meant "get out." Stopgap measures have become less and less effective. By the early twenty-first century, even those dairy farmers who switched to organic milk production in the 1990s faced the same catastrophic losses that conventional farmers had faced a few decades ago.[27] The sheer unsustainability of modern farming seems evident now in everything from farm bankruptcies to Lake Champlain's persistent problems with farm run-off pollution.

The human impact, too, has begun to be obvious. Many of the Vermont dairy farms that remain in operation have come to rely almost completely on workers from Mexico and Central America—workers who are often undocumented and who have few protections from exploitation.[28] The price of land now makes it nearly impossible for new people to acquire farms. And in the years following the COVID epidemic of 2020, food insecurity—hunger, in the language of a more brutal time—has been at an all-time high.[29] Naturally, there are many organizations working to solve these problems: to provide food to the hungry, support systems for immigrants, subsidized land for new farmers. But where now can one find that treasured commodity Professor Peach spoke about in 1929: "a home from which a man can neither be starved nor frozen as long as a crop will grow and wood burn"?

Returning to the Hills

In the early years of the twenty-first century, the threats have continued to grow. Climate change has already caused severe damage to the farms of Vermont as unprecedented rainfall events brought three catastrophic floods in the single year between July 2023 and July 2024, flooding the capital city and causing inestimable losses to Vermont's farms, now almost all located in narrow river valleys and low-lying plains. Perhaps the situation is now dire enough to prompt a reassessment of the hill farms, and even of the trajectory of farming as a whole.

The phrase "hill farm" has never meant simply farms on slopes: it has always referred to scale, method, and purpose. In that sense, there are many farmers these days who are making valiant efforts to create—or to recreate—what they may or may not recognize as a hill farm legacy. In many ways, they are the successors (sometimes literally the children) of the back-to-the-land growers of the 1960s and 1970s. But to a surprising degree their approach might also sound familiar to a hill farmer from the 1930s. Under the pressure of multiple environmental crises, these farmers are probably much more aware of the fragility of their land than were the hill farmers of the past, and they certainly have more specialized tools at their disposal. But they share some of the tenacity and creativity of the hill farmers from a century ago.

One farmer near Burlington, Vermont, for example, breeds a landrace of hardy cattle closely adapted to her low-lying clay soils. As her herds graze, they increase the depth of her soil and the quality of its grass cover. Hill farmers did not know about modern intensive mob grazing or carbon sequestration, but they would have recognized what Bread and Butter Farm calls their "grandma-centric" herd, now browsing the woods and shrubs at the edge of the fields, like hillside cattle in days long past. Making use of woodland grazing and foraging was a defining feature of hill farms a hundred years ago. Making use of new models—silvopasture, agroforestry, permaculture—some farmers are learning to work productively in similar ways within second-growth wooded landscapes.

Other farms have sprouted up in locations that now seem almost impossibly challenging—but were commonly farmed back in the 1930s. Small Axe Farm in Barnet, Vermont, grows vegetables at 1,500 feet of elevation, on a hillside so steep they cannot use most power tools. "We were able to afford this land because it's steep," the two farmers reported in 2023.[30] To be sure, the old hill farmers knew nothing about the kind of no-till farming practices of Small Axe Farm or their state-of-the-art solar power array; but they knew a lot about using hand tools on steep slopes, and about making do with less than ideal farmlands. Small Axe vegetable fields slope intentionally southward, as was "best practice" long ago. And like the farmers of Jamaica in 1930, Small Axe Farm initially produced for "home use"; it was first established as a homestead before branching into commercial farming. The farmers cultivate only a very small portion of their land—a single acre of raised beds "hard-won" from old pasture a little at a time, by hand. The rest of their forty-five acres they leave in pasture and woodland—"to build soil," as Robert Frost put it in 1932. As the farmers wrote on their Instagram account in fall 2024, "we were spared the worst of the flooding that happened in the valleys. Our roads were destroyed, but our farm was not."[31]

Perhaps the most intriguing feature of the new farming systems is their preference for overlapping layers of productivity. Today that kind of "systems thinking" is understood to promote resilience and flexibility in the face of environmental and social challenges—but it is also very reminiscent of the old-fashioned farms of Jamaica. On one farm, for example (to be sure, this one is in New Hampshire), sheep and cattle graze overgrown fields and woods on what the farmers describe as a "wooded, hilly, rock-strewn" piece of land. Pigs root in the underbrush; firewood is cut strategically, to free space around young maples and expand the sugarbush. The wood is used to fire the farm's wood-burning maple evaporator. The farm sells maple syrup, lamb, beef, pork, and firewood—and vegetables grown on six acres of their most level land.

Nearly every aspect of this operation would have been condemned by the agricultural specialists of the 1930s, from the rocky, slopy land itself to the diversified crops the farm produces. In fact, Kearsarge Gore Farm even resembles the family arrangements of the farmers of Jamaica in 1930: the first-generation founders of the farm work with their son and his partner. (Presumably the young people would be described as "staying home and doing nothing" by a modern eugenicist.) Even what they like about farming sounds familiar: "I appreciate the diversity of work, it's not the same thing every day."[32] This is the kind of farming the eugenicists encountered in Jamaica—even to the very percentages of land used for gardens, hay, and woodland. It is precisely the kind of farming the experts have always predicted would fail.

And indeed, these kinds of farmers *would* fail if they did not build diversity into their income streams, much as Jamaica farmers once did. On these new farms, they host summer camps and cooking schools; they market their vegetables and meat through farmers' markets and CSAs and online home delivery services; and they make and sell what are now called "value-added" products: beer and jam; sauerkraut and ice cream. They hold all kinds of off-farm jobs: not always on the town roads anymore, but sometimes even that. And many of them—like a previous generation of farmers who lived in the shadow of industrializing agriculture—make their homes high up in the hills, surrounded by second-growth forest, on the only land they can afford.

It is my hope that my stories from Jamaica might in some way offer aid to the efforts of these farmers in the hard times they face, and perhaps even help us all to think a bit more clearly about how to live within the limits our land imposes on us. We cannot return to the farming landscape of the past. We need our forests, now more than ever as the climate catastrophe unfolds. But we need farms, too. As farming in more prosperous regions faces increasingly dire threats from fire and flood, who knows now whether we might someday depend on hill farms again.

Notes

Introduction

1 "Jamaica Local Eugenics Survey Completed," *Brattleboro Daily Reformer*, December 26, 1930.
2 An important first example was the work of Charles O. Gill and Gifford Pinchot, *The Country Church: The Decline of its Influence and the Remedy* (New York: Macmillan, 1913). These surveys were what historian Hal Barron had in mind back in the 1980s, when he proposed that social historians investigate the countless surveys and questionnaires generated by the new social sciences of the early twentieth century. Hal S. Barron, "Rural Social Surveys," *Agricultural History* vol. 58, no. 2 (April 1984): 113–17.
3 The foundational work on the Vermont eugenics movement is Nancy Gallagher's *Breeding Better Vermonters* (1999), an absorbing and scholarly analysis of the ideas that informed Henry Perkins' career as a eugenicist. An important recent work considers the eugenics movement within a broader social and political environment that fostered the acceptance of eugenical policies and institutions in Vermont: Mercedes De Guardiola, *Vermont for the Vermonters: The History of Eugenics in the Green Mountain State* (2023).
4 Some have asserted that the Eugenics Survey targeted indigenous Abenaki communities. In the records I have examined, evidence for that assertion is sparse. The Migration Study records reveal occasional uses of the categories "Indian" and "Abenaki," but they do not focus on that population. The question is not purely "academic;" it has become part of a larger debate over state recognition of the claims of several self-identified Abenaki groups. Challenged by the leaders of the Quebec Abenaki community of Odanak to provide proof of their Indigenous ancestry, some members of the Vermont communities have suggested that all proof was lost when their families were targeted by the Eugenics Survey. https://www.vermontpublic.org/podcast/brave-little-state/2023-10-19/vermont-recognized-tribes-canada-abenaki-first-nations-odanak-wolinak.
5 That is not to say that there were no people in town of non-Yankee descent. In the notes of the eugenicists, fifteen percent of the households interviewed were reported to include at least one person of a different background. Eugenics Survey of Vermont,

"Selective Migration from Three Rural Vermont Towns," *Fifth Annual Report of the Eugenics Survey of Vermont* (Burlington: University of Vermont, 1931), 43.
6 Perkins famously cited the supposedly poor medical records of Vermont's rural, largely "Yankee" draftees during World War I as the catalyst for his interest in eugenics studies. Gallagher, *Breeding Better Vermonters*, 39–40.
7 E. R. Pember, "Our Hill Farms," *Eighth Vermont Agricultural Report* (Montpelier, VT: Watchman and Journal Press, 1884), 362.
8 Herbert Agar, "Introduction," *Free America* vol. 1, no. 1 (January 1937): 1.
9 In his dissertation, global historian Christopher Harris provides an extraordinarily wide-ranging and deeply rooted global framework for this process of change, making insightful use of the groundbreaking work of economist Esther Boserup and anthropologist Robert McC. Netting. See especially chapter 2, "Why the Modern World Hates Small Farmers." Christopher Harris, "The Road Less Traveled By: Rural Northern New England in Global Perspective, 1815–1960," PhD dissertation, Northeastern University, 2007.
10 To make a single comparison, several of the great social historians of the 1960s and 1970s utilized the records of the Catholic Inquisition to reconstruct the social and intellectual worlds of medieval people caught in its grip. See, for example, Carlo Ginzburg, *The Cheese and the Worms* (1976); Emmanuel LeRoy Ladurie, *Montaillou: The Promised Land of Error* (1975); Natalie Zemon Davis, *The Return of Martin Guerre* (1983).
11 The documents, along with all the materials from the Eugenics Survey office, were moved to the University of Vermont's Fleming Museum when the eugenics office closed in 1936. From there they were moved to the Vermont State Archives. They are open to the public, as is the mandate for all state archives—with the requirement that the names of the subjects be kept private. In this book I have given pseudonyms to everyone whose name appears in eugenics records, and to all others whose names I have linked to those documents. The Migration Study documents are held at the Vermont State Archives and Records Administration, hereafter VSARA. Series PRA005: Eugenics Survey of Vermont and the Vermont Commission on Country Life Records, 1925–1956. Migration Study: Containers PRA-00014 and PRA-00015.
12 The evolution of progressive and Country Life agricultural reform into New Deal agricultural and environmental programs is handled in different and complementary ways by Sara Gregg in *Managing the Mountains: Land Use Planning, the New Deal, and the Creation of a Federal Landscape in Appalachia* (2013); Sarah Phillips in *This Land, This Nation: Conservation, Rural America, and the New Deal* (2007); and Jess Gilbert in *Planning Democracy: Agrarian Intellectuals and the Intended New Deal* (2015).
13 My choice to focus on a single town and moment in time clearly owes a great deal to the foundational work of Hal Barron, *Those Who Stayed Behind: Rural Society in Nineteenth-Century New England* (1984), which explores rural life and agriculture in the town of Chelsea, Vermont, in the nineteenth century. Other similar important local studies of related matters are: Grey Osterud, "Farm Crisis and Rural Revitalization in South-Central New York during the Early Twentieth Century," *Agricultural History* vol. 84, no. 2 (2010); and Paul Searls, *Repeopling Vermont: The Paradox of Development in the Twentieth Century* (2019).

14 Brian Donahue, *The Great Meadow: Farmers and the Land in Colonial Concord* (New Haven: Yale University Press, 2007). Donahue makes a powerful case that eighteenth-century Concord farmers' practices were well adapted to the opportunities and constraints of their land, and that they were finally defeated not by bad farming but by the development of a capitalist agricultural system in the nineteenth century. Perhaps equally important, Donahue's work offers an eye-opening analysis of just how much expertise and care went into successful farming in Concord.

15 Sarah Phillips, *This Land, This Nation: Conservation, Rural America, and the New Deal* (Cambridge: Cambridge University Press, 2007), 40. Phillips refers specifically here to the concept of the "marginal" and "submarginal" farm.

16 I did not conduct oral histories in Jamaica, but there are excellent examples of other historians' use of such an approach to inquire into the feelings and values of rural people. Grey Osterud's illuminating study of farm communities in upstate New York's Nanticoke Valley offers an intriguing contrast with my story of Jamaica, and her use of oral histories allowed her to uncover much about the internal life of the community. Grey Osterud, "Farm Crisis and Rural Revitalization in South-Central New York during the Early Twentieth Century," *Agricultural History* vol. 84, no. 2 (2010). Paul Searls makes excellent use of oral histories to examine similar questions in *Repeopling Vermont: The Paradox of Development in the Twentieth Century* (2019), which describes how the town of Landgrove—just two towns north of Jamaica—was transformed by newcomers and progressive reforms in the early twentieth century.

17 In a striking testimony to how region and class can overlap with each other, that is exactly the same phrase I heard about my own grandparents' generation as I was growing up in south Texas.

18 The story is recounted in detail by Sara Gregg in *Managing the Mountains*. Aiken himself tells his own version of the story in "Not So Submarginal," *Speaking from Vermont* (New York: Frederick Stokes and Co., 1938).

19 She attributed the idea to her grandfather. Dorothy Canfield Fisher, "Vermonters," *Vermont, a Guide to the Green Mountain State* (Boston: Houghton Mifflin, 1937), 3.

20 The poem adopts the forms and characters of Virgil's Eclogues and considers many things both political and artistic in addition to farming. Robert Frost, "Build Soil: A Political Pastoral," *A Further Range* (New York: Holt and Co., 1936).

Chapter 1

1 Harold Fisher Wilson, for example, wrote that "inaccessible hill farms were carved from the forest with little thought of their potential disadvantages." *The Hill Country of Northern New England* (New York: Columbia University Press, 1936), 16. Lewis Stilwell suggested that the earlier settlers had been the victims of a "bitter practical joke," believing in Vermont as "the land of agricultural promise" because "the rich legacy of the slaughtered forests" had provided a short burst of fertility, an "aboriginal goodness" that was then "used up or washed away." *Migration from Vermont* (Montpelier, VT: Vermont Historical Society, 1948), 71.

2 A major influence has been Brian Donahue's *The Great Meadow: Farmers and the Land in Colonial Concord* (2004). Donahue's work calls into question earlier assumptions

about the destructiveness of colonial agriculture and examines how and why early farmers reached the limits of their land's ability to support them. Donahue finds among Concord's farmers a degree of sophisticated understanding of the land's multiple possibilities that is nearly the opposite of earlier historians' assessments.

3 Environmental historian Richard Judd provides an excellent summary of the history of the northern migration in *Second Nature*. As he writes, "Land secured parental authority, kept the extended family in one place, provided security in old age, and buttressed the community." Richard Judd, *Second Nature: An Environmental History of New England* (Amherst: University of Massachusetts Press, 2014), 78.

4 Judd, *Second Nature*, 91.

5 Peter B. Mires, "The Importance of Aspect to Historic Farmstead Site Location in the Green Mountains of Vermont," *Historical Archeology* vol. 27, no. 4 (1993): 88–89.

6 Samuel Williams, *The Natural and Civil History of Vermont* (Burlington, VT: Samuel Mills, 1809, second edition), 25.

7 E. R. Pember, "Our Hill Farms," in *Eighth Vermont Agricultural Report by the Board of Agriculture, 1883–1884* (Montpelier, VT: Watchman and Journal Press, 1884), 363.

8 Mabel Cook Coolidge, *The History of Petersham, Massachusetts* (Holyoke, MA: Powell Press, 1948), 25–26.

9 This innovation in land use had its roots in "soldier townships" like Petersham, where the Massachusetts government substituted gridded lots distributed to individuals for allotments that followed topography. Richard Judd suggests that it was Benning Wentworth, the chief force behind the aggressive surveying and settlement of the land that would become Vermont, who was the "grandest innovator" of the grid system. *Second Nature*, 78.

10 In these early transactions, the actual boundaries within the roughly 100-acre lots were not specified: the piece of land bought was simply an abstract portion of the division lot—half or a quarter "to be divided equally as to quantity and quality." Elisha bought all of Lot 30 and half of Lot 210, roughly 150 acres in two plots that lay in the southeastern part of town.

11 In 1793, Stephen Chase bought half of Lot 103, a little to the west. He added the adjoining eastern half of Lot 104 in 1800 to total about 100 acres in all. Peter Chase bought a quarter of Lot 9—around twenty-five acres—adjacent to and just north of Elisha's lot, to which he added another forty-eight acres in 1800, part of nearby Lot 11 on what was probably adjacent land. In 1806, while he consolidated his holdings on the main farm, Peter also bought a half of Lot 102, a 100-acre lot a little farther west, contiguous with Stephen's lot. In 1822, he traded two different pieces of Lot 101 to consolidate his holdings there.

12 Elisha sold to Peter forty acres from the northern part of his property, land that bordered his brother's farm. In 1815, Elisha traded two small plots with adjoining neighbor Richard Smith, and in 1817 Peter traded with his neighbor Ezra Ober, probably to make more workable boundaries between their farms. In 1824, Elisha sold "the farm on which I now live," including sixty acres and two houses.

13 The first farms were located on the flood plains of the West River and its tributaries, but it quickly became obvious how dangerous that strategy would be. The bench lands (stepped terraces shaped by the geological forces of rivers and streams as they wore down their beds) proved safer, and that is where most early farms were located. Mark

Worthen, *Hometown Jamaica: A Pictorial History of a Vermont Village* (Brattleboro, VT: Griswold Offset Printing, Inc., 1976), 18.

14 Probably the lot owned by Richard Smith, who appeared next to Elisha Chase on the 1810 census. At some point the Chases acquired it, and it became Martin's when Peter Chase bequeathed it to him in 1851.

15 "Grand list" is the term used in Vermont (and Connecticut) for the property and poll taxes levied by the towns. The records are held in the offices of each town's Town Clerk. For Jamaica, the 1790 grand list is the first surviving record of the town's poll and property taxes, but no Jamaica grand lists survive between 1793 and 1806.

16 Estimates from other places in Vermont range from one to four acres a year per adult man. See, for example, *A Narrative of the Sufferings of Seth Hubbell & Family, in his Beginning a Settlement in the Town of Wolcott, in the State of Vermont* (Danville, VT: E. and W. Eaton, 1829). Historians of the Harvest Forest land estimated based on historical records that a man working on his own might clear from one to three acres per year. Hugh M. Raup and Reynold E. Carlson, *The History of Land Use in the Harvard Forest* (Petersham, MA: Harvard Forest, 1941), 2. A tax list from the town of Peacham (in northeastern Vermont) reveals that over a decade after the first division of the land the town's farms still averaged just five acres each of cleared tillage. Ernest L. Bogart, *Peacham: The Story of a Vermont Hill Town* (Montpelier: Vermont Historical Society, 1948), 73.

17 The Harvard Forest was established in 1907 as a laboratory for students in Harvard's School of Forestry. The museum affiliated with the Forest created twenty-three diorama exhibits depicting aspects of forest change over time. Seven of the twenty-three comprise the "Landscape History of Central New England." https://harvardforest.fas.harvard.edu.

18 Warren E. Booker, *Historical Notes: Jamaica* (Brattleboro, VT: Hildreth Co., 1940), 73.

19 This man had grown up working in a family that still made potash in the nineteenth century. Walter Needham and June Barrows Mussey, *A Book of Country Things* (Brattleboro, VT: Stephen Greene Press, 1965), 114–115.

20 James H. Phelps, *Collections Relating to the History and Inhabitants of the Town of Townshend, Vermont* (Brattleboro, VT: 1877), 13, 145.

21 Williams, *Natural and Civil History*, 25.

22 Zadock Thompson, *Gazetteer of the State of Vermont* (Montpelier, VT: E. P. Walton, Printer, 1824), 161.

23 Thompson was a self-taught first-generation Vermont scholar who compiled the first histories and natural histories of the state. His work was optimistic, but not notably exaggerated. See J. Kevin Graffagnino, "Zadock Thompson and the Story of Vermont," *Vermont History* vol. 46, no. 2 (Spring 1978): 77–99.

24 John Hayward, *Gazetteer of Vermont* (Boston: Tappan, Whittemore and Mason, 1849), 97, 120. Describing Stratton's soil as "cold" and Jamaica's as "warm" appears to have been a way of distinguishing clay soils from sandy soils. As one contemporary expert explained, "All sands, are hot and dry—all clays, cold and wet." Thomas Green Fessenden, *Complete Farmer and Rural Economist* (Boston: Russell, Odiorne, and Co., 1835 second edition), 11.

25 Hamilton Child, *Gazetteer and Business Directory of Windham County, Vermont* (Syracuse, NY: 1884), 222.

26 James Dean, *Alphabetical Atlas of Vermont* (Montpelier, VT: Samuel Goss 1808), 6.
27 Thompson, *Vermont Gazetteer*, 14.
28 Brian Donahue's account of the role of cattle in early Concord is instructive for all students of New England agriculture: "The cattle lead us down both the market highways and the local cartways of this economy and always bring us back to the ecological bottom of the matter, which was the grass growing in the meadows." *The Great Meadow*, 12.
29 One author in the 1870s estimated that cows might be milked for around ten years before being slaughtered for meat. (Today industry standard life for dairy cows is around five years, and their meat is often sold as low-grade ground beef for fast food.). Albert Chapman, "What Breed of Cattle Shall We Raise?" *Annual Report of the Vermont Board of Agriculture* (Montpelier, VT: Board of Agriculture, 1872), 172.
30 This calculation comes from the grand list for 1830, which listed 1,513 cattle, and the 1830 census, which listed 216 heads of households.
31 Virginia DeJohn Anderson, *Creatures of Empire: How Domestic Animals Transformed Early America* (2006) offers a useful discussion of New England's increasingly widespread cattle trade in the seventeenth century.
32 Dean, *Alphabetical Atlas*, 7.
33 Thompson, *Gazetteer*, 15.
34 Thompson *Gazetteer*, 311; John Hayward, *New-England Gazetteer* (Concord, NH: Israel S. Boyd and William White, 1839).
35 Animals were driven to Massachusetts markets from northern New England on foot, by road and ferry or bridge. No turnpike ran directly through Jamaica, but the Stratton turnpike ran through Wardsboro just to its south, offering a direct road to Brattleboro and beyond. Stilwell, *Migration from Vermont*, 102–3.
36 Thompson, *Gazetteer*, 14.
37 Chase's flock was in the top 20 percent of sheep flocks in 1829.
38 The compilers of the booklet *A Statistical View of the Number of Sheep* (yes, there was a book!) counted 1,099,011 sheep in the state that year, and the number had not yet peaked. C. Benton and S. F. Barry, compilers, *A Statistical View of the Number of Sheep in the Several Towns and Counties of Maine, New Hampshire, Vermont . . . in 1836* (Cambridge, MA: Folson, Wells and Thurston, 1837), 34 and 122. The 1840 number is reported in Isaac Newton, *Report of the Commissioner of Agriculture for the Year 1864* (Washington: Government Printing Office, 1865), 253, as cited in Robert F. Balivet, "The Vermont Sheep Industry: 1811–1880," *Vermont History* vol. 33, no. 1 (January 1965): 247.
39 Harold Fisher Wilson, for example, suggested that the "hill country of Northern New England was peculiarly adapted to the raising of sheep" because of its "rock-strewn fields and pastures," and repeated the old joke about the sheep's sharp noses fitting in between the stones. "The Rise and Decline of the Sheep Industry," 13.
40 The 1836 sheep tally listed only nine towns in Vermont with fewer than a thousand sheep. All nine were mountain towns, and three of those towns bordered Jamaica on the west: Somerset, Stratton, and Winhall. D. K. Young, *The History of Stratton, Vermont to the End of the Twentieth Century* (Stratton, VT: Town of Stratton, 2001), 45. One gazetteer explained that the mountain town of Searsburg, for example, was simply "too elevated on the Green Mountains for either cultivation, population, or wool growing." Hayward, *Gazetteer of Vermont*, 113.

41 Some Champlain Valley farmers, for example, had accrued both cash reserves and familiarity with volatile markets by their experience with commercial wheat farming, only recently brought to a sudden end with the advent of the destructive wheat rust disease. It was this business that gave them the commercial experience and start-up capital to get involved with wool growing. P. Jeffrey Potash, *Vermont's Burned-over District: Patterns of Community Development and Religious Activity, 1761–1850* (Brooklyn, NY: Carlson Publishing, 1991).
42 The *Statistical View of the Number of Sheep* (1837) reported that wool was commonly sheared from mixed-breed sheep. The best-quality wool came from "half to full blood Merino." In Windham County the wool was generally "one half to three quarters blood." *Statistical View*, 23, 26.
43 Stilwell, *Migration from Vermont*, 172–3.
44 This process is described in Potash, *Vermont's Burned-Over District*, 104–116. Potash links the social disruption of the sheep craze with the outbreak of the Second Great Awakening in the valley. "The promise for some residents shone more brightly than ever before, for others the promise flickered, and for increasing numbers, the promise was extinguished." 116. Jamaica, in contrast, followed the pattern of the more mountainous and remote towns in Vermont. It experienced relatively little of the commotion associated with the mainstream evangelicalism of the Second Great Awakening, but instead, was drawn to the more "fringe" Millerite movement of the 1840s.
45 That was not the end of it: by 1840, Addison County reported 260,000 sheep, and Shoreham 41,118. Balivet, "Vermont Sheep Industry," 246.
46 It is also not a reliable number. The published numbers in *A Statistical View of the Number of Sheep* are sometimes seriously at odds with the numbers recorded on the town grand lists, despite the authors' claim to have obtained their information from "persons holding official stations." This figure of 3,863 is significantly higher than the counts in the Jamaica grand lists. Still, the numbers seem to be accurate enough to allow for a general comparison with other towns.
47 Jeffrey Potash found that even in Addison County there were dramatic differences between the lowland towns and the hill towns on the eastern side of the county. In the hill towns, the numbers of sheep were markedly lower, and both cattle and human populations were stable—as they were in Jamaica. Potash, *Vermont's Burned-Over District*, 115.
48 In 1836, for example, Samuel Fessenden held a flock of 91 sheep, George Howard a flock of 90, and Hammond Livermore a flock of 101.
49 Potash, *Vermont's Burned-Over District*, 117.
50 Newton, *Report of the Commissioner of Agriculture*, as cited in Balivet, "Vermont Sheep Industry," 243 and 247. 1,681,819 in 1840; 1,014,122 in 1850; 752,201 in 1860.
51 In some communities, farmers found that falling prices could be made up for to some degree by breeding more productive sheep. Hal Barron describes how increases in productivity allowed Chelsea farmers to keep farming sheep into the 1890s. Barron, *Those Who Stayed*, 64. Harold Fisher Wilson reports that in the 1850s, although prices rose a little, it was not enough to cover costs in northern New England: "For instance, the yearly cost of keeping a sheep in Windham County, Vermont, in 1851, was from $1.25 to $1.50, while the average fleece weighed three pounds. Wool in that year sold for forty cents a pound, which amounted to an income of only $1.20 per

sheep,—less than the cost of production." Wilson, "Rise and Decline of the Sheep Industry," 21.

52 In 1843, for example, just 50 percent of heads of households owned any sheep at all; 22 percent owned between one and ten sheep; 24 percent owned between eleven and forty-nine; and 4 percent (fifteen households) owned fifty or more sheep. I did not place these numbers in the table because they come from the grand lists and are not exactly comparable to the census-derived numbers.

53 Robbins's farm straddled the line between Jamaica and Windham, with his house and barn in Windham and his land and other buildings in Jamaica. His farm was counted in Jamaica in 1860, in Windham in 1870 and 1880.

54 Estimates from nineteenth-century sheep experts suggest that Robbins's large flock might have required anywhere between 80 and 250 acres to provide adequately for both summer grazing and winter feeding. The Revolutionary leader Robert R. Livingston, an early promoter of Merinos in the United States, estimated that farms in Pennsylvania could support an average of one to two sheep per acre including both hay for winter feed and summer grazing. The author of *The American Shepherd* suggested in 1845 a carrying capacity of a hundred sheep for thirty-five acres including both winter and summer feeding: a little under three sheep per acre. In 1849 *The Wool Grower and Magazine of Agriculture and Horticulture* reprinted a note from *The Iowa Farmer*: "It is said by some agricultural writers, that it takes one acre of land to sustain five sheep during summer and winter. We, however, will calculate one acre to *three* sheep." Robert R. Livingston, *Essay on Sheep* (New York: T. and J. Swords, 1809), 110; as cited in Steven Stoll, *Larding the Lean Earth* (New York: Hill and Wang, 2002), 113. Luke A. Morrell, *The American Shepherd* (New York: Harper and Brothers, 1845), 208. *The Wool Grower and Magazine of Agriculture and Horticulture* vol. 1, no. 1 (March 1849): 11.

55 Brian Donahue, "Another Look from Sanderson's Farm: A Perspective on New England Environmental History and Conservation," *Environmental History* vol. 12, no. 1 (Jan. 2007): 18–19.

56 Potash, *Vermont's Burned-Over District*, 106.

57 Wilson, "Rise and Decline of the Sheep Industry," 15.

58 The practice of grazing sheep and cattle on the same pastures seems to have been well established. Many years later, Vermont politician George Aiken wrote that it had formerly been the general practice of "thrifty" hill farmers in southeastern Vermont to rotate sheep through all their grazing land every few years to keep the pastures from getting "frowsy." George Aiken, *Speaking from Vermont* (Boston: Frederick Stokes Co., 1938), 15.

59 The agricultural censuses defined "improved" land as "cleared land used for grazing, grass, or tillage, or currently held fallow." "Unimproved" land specifically included woodlots or land whose timber was used in connection with the farm. Numbers were not always consistently reported or calculated, and category definitions changed over time. But for general comparison between Jamaica and the state as a whole, the numbers should be adequate.

60 Marsh argued that deforestation had led to desertification in the Mediterranean lands of the classical past, and that the same thing could happen in New England and the rest of the United States. George Perkins Marsh, *Man and Nature, or, Physical*

Geography as Modified by Human Action (London: Sampson Low, Son, and Marston, 1864.)
61 In 1860, Chelsea averaged thirty sheep per farm, in comparison with Jamaica's average of eleven. Barron, *Those Who Stayed*, 64. It does not seem to have been a matter of superior farming. In 1880, the average fleece weight in Jamaica was five pounds—almost exactly the same as Chelsea's. That year, the state average weight of wool was somewhat higher (5.8 pounds), but the national average was still only 4.4 pounds.
62 Barron, *Those Who Stayed*, 58–59.
63 In fact, two of those six farms held heifers that may have been about to become milk producers.
64 94 percent of farmers in 1880.
65 Jamaica averaged 272 pounds of butter per household versus the statewide average of 711 pounds.
66 The critical labor shortage of farm labor in Chelsea, as elsewhere, was rooted in outmigration.
67 For an excellent summary of another town's struggle to adapt to the demands of dairying, see Shepard B. Clough and Lorna Quimby, "Peacham, Vermont: Fifty Years of Economic and Social Change, 1929–1979," *Vermont History* vol. 50, no. 1 (Winter 1983).
68 The agricultural census manuscript schedules are available for 1850, 1860, 1870, and 1880. All four censuses report numbers of animals per farm, but the 1880 is by far the most detailed and provides the basis for much of this discussion. Yields cannot be derived from the earlier censuses because they recorded bushels of crops without recording acreage.
69 In 1880, 93 percent of Jamaica farms grew potatoes. They averaged about 73 bushels per acre as compared to the state average of 115. It would be more accurate to average yields only for those who actually grew potatoes, but the statewide averages can be computed only for all farms, so this number offers a fairer comparison. The yield calculated only for those in Jamaica who grew potatoes was about seventy-eight bushels per acre.
70 Corn was close to the extreme limit of its range in parts of the hills. Jamaica farmers collectively produced an average of twenty-nine bushels of corn per acre, about 20 percent lower than the statewide average of thirty-seven bushels. Jamaica farmers reported corn yields ranging from eight to sixty-seven bushels per acre—that is, from less than a quarter of the average state yield to nearly twice the average state yield. In *A Long, Deep Furrow*, Russell estimated that corn yields in early nineteenth-century New England varied from a very low "8 bushels an acre on old poor land"—the lowest yield in Jamaica in 1880—"to 25–30 bushels in well fertilized places." By mid-century, an average corn yield was thirty bushels an acre. Jamaica's average in 1880 was still twenty-nine. Howard S. Russell, *A Long, Deep Furrow: Three Centuries of Farming in New England* (Hanover, NH: University Press of New England, 1976), 278, 367.
71 One 1916 source reported that Vermont farmers in Windsor County (north of Jamaica) spread manure only on their potato and corn crops, as part of a rotation which required manuring of a field only once every six years or so: one year of manured corn or potatoes followed by oats and then by grass for four years or more. J. A. Kerr and Grove B. Jones, *Soil Survey of Windsor County, Vermont* (Washington, DC: Government Printing Office, 1918), 11.

72 Judd, *Second Nature*, 82. As late as 1880, artificial fertilizer was not commonly used in the hills: only 21 Jamaica farmers (out of 210) reported they had spent anything at all for it that year. Farmers were not asked about lime, nor about leguminous pasture crops that might have assisted them in maintaining their soil's fertility.

73 In 1880, 92 percent of farms held some wooded acreage, on average fifty acres as compared to the state average of forty-two.

74 The town average was twelve cords of wood per farm, while the state average was just under twenty cords.

75 Vermont's average production of maple sugar in 1880 was 317 pounds per farm (plus 3.6 gallons of the far less widely used maple "molasses"). In Jamaica, that figure was 15 percent less—just 270 pounds.

76 It is possible, for example, that Jamaica had few maple trees because they had been cut down and had not yet re-grown in large numbers. But the 1884 Child's *Gazetteer* listed thirty-nine farms with a specialty in maple sugar, and those farms reported numbers of maple trees ranging from a low of two hundred to a high of three thousand—so at least on those farms, the trees do not seem to have been scarce.

77 18,240 trees tapped out of 28,266. "Statistical Abstract," *Eight Annual Report of the Commissioner of Agriculture of the State of Vermont* (St. Albans, VT: St. Albans Messenger Co, 1916), 47–56.

78 Counting twenty-five to fifty gallons of syrup as the equivalent of around two to four hundred pounds of sugar. Needham, *Book of Country Things*, 40–41.

79 That was the same labor shortage, of course, that slowed the adoption of commercial dairy farming in places like Chelsea, as Barron points out.

80 A study of thirteen hill towns conducted in 1929 would report that maple yields in the hill towns were determined mainly by the intensity of cultivation, and particularly by the amount of time household members—often working off-farm jobs as well—were able to devote to farming. Lemuel J. Peet, "Problems of Land Utilization in the Hill Towns of Vermont" (master's thesis, University of Vermont, 1930), 107–114.

81 A 1916 soil survey of nearby Windsor County suggested that farmers in the hill towns were accustomed to grazing at least some of their animals on wooded land. "Mowing land is seldom pastured, as there are large fields of wood land or stony land suitable for pasture." Kerr and Jones, *Soil Survey of Windsor County*, 11.

82 "Vermont Cattle by Railroad to Boston," *Scientific American* vol. 4, no. 16 (January 6, 1849).

83 *Vermont Watchman*, January 23, 1851, as quoted in Stillwell, *Migration from Vermont*, 221.

84 Henry David Thoreau, *Walden, or Life in the Woods* (London: Vintage, 2017), 109.

85 Jerseys were the preeminent high-fat dairy cattle prized by farmers who were producing rich cream for butter. "Dutch" referred to the high-production Holsteins who would eventually come to populate nearly every dairy farm in the state. Durhams were Shorthorns, dual purpose animals that were sometimes bred to specialize in either meat or dairy.

86 Axel Anderson and Florence M. Woodard, "Agricultural Vermont," *Economic Geography* vol. 8, no. 1 (January 1932): 31.

87 Thoreau, *Walden*, 109.

88 They reported 97 sheep in the 1865 grand list, 92 sheep in the 1870 census, and 160 sheep in the 1880 census.

89 One home economics study calculated that Vermont farm families required from one to two hundred pounds of potatoes per person per year, which would work out to around eighteen bushels for the Howes' family of three adults and three children. They might also consume about ten bushels of apples over the winter. Dorothy Emery, "A Food Budget for Vermont Farm Families," *Vermont Agricultural Experiment Station Bulletin 393* (Burlington: Vermont Experiment Station, 1935), 8. Another source reports that households retained roughly ten gallons of syrup and fifty pounds of maple sugar for their own use. Anderson and Woodard, "Agricultural Vermont," 39. So in all these categories, they were aiming to market the surplus.

90 Only some categories can be compared, since there is no measured yield recorded for eggs or butter, for example.

91 They got ninety bushels on three acres of land. Only 32 farmers raised buckwheat in Jamaica (171 raised corn, in comparison), and most raised far less. Buckwheat could be fed green or dried to cattle and other animals and could be used as a catch crop, to fill a gap after harvesting or to provide protection for slow-growing crops. Thomas Bjorkman writes in a blog post that "Buckwheat's low input requirements made it the mainstay on hill farms with low soil fertility and weak finances." Bjorkman reports that in the early twentieth century, buckwheat became associated with poverty and was avoided by farmers in New York, with some going so far as to hide their buckwheat on fields not visible from passing roads. *Cornell Field Crops*, blogs.cornell.edu/whatscropping up/2016/05/26/buckwheat-a-historic-crop-with-modern-opportunities.

92 One man born in southeastern Vermont conveyed the complexity of his skill to his grandson, who lived with him around the turn of the century on his farm, learning the skills his grandfather had learned as a child back in the middle of the nineteenth century. His account showcases an almost uncanny set of skills from making tools to diagnosing and treating animal illnesses. Needham, Mussey, *Book of Country Things*.

93 As one Old Home Day orator described it, the debt was difficult to repay after the war and was made even more difficult by the destruction caused by a major flood in 1869. As quoted in Booker, *Historical Notes*, 115 and 117.

94 He returned to Vermont after a few years, but not to Jamaica, settling in the nearby mountain town of Somerset, where he married, had children, and established a lumber business. (Originally Daniel Franklin Chase, he called himself Franklin, or Frank, in later years.)

95 Booker, *Historical Notes*, 70–71.

96 Child, *Gazetteer Windham County*, 421–33.

97 The obituary of the man listed in the 1900 census as the owner of the farm refers to the farm as the "Arad T. Wood place." *Brattleboro Daily Reformer*, January 15, 1914. I have not named him here because he was one of the people interviewed by the eugenicists in 1930.

Chapter 2

1 Most agricultural historians' knowledge of this pattern of rural population decline has been shaped by Hal Barron's *Those Who Stayed Behind* (1984), an exploration of the decline of the town of Chelsea, Vermont, in the late nineteenth century.

2 Agrarianism was such a powerful ideology that even groups explicitly disqualified from the American agrarian dream often attempted to use it for their own purposes; see, for example, the Tuskegee Institute's educational goals for Black farmers. Dona Brown, *Back to the Land: The Enduring Dream of Self-Sufficiency in Twentieth-Century America* (Madison: University of Wisconsin, 2011), 49–50.

3 A good definition of agrarianism was provided by agricultural historian Clifford B. Anderson back in the 1960s. First, farms are the "primary producers" upon which all other productivity rests; second, farming is not just a trade but a "way of life" morally superior to urban trades because the farm is close to nature and God; and third, the United States should remain a nation of small farms to avoid political and moral decline. "The Metamorphosis of American Agrarian Idealism in the 1920s and 1930s," *Agricultural History* 35, no. 4 (October 1961): 182.

4 Such early writers as Rose Terry Cooke, Elizabeth Stuart Phelps, and Sarah Orne Jewett acknowledged poverty and depopulation in the rural north but portrayed New England's remaining rural people—especially the women—as upright, courageous, and wise. Later writers tended to focus on the bleak physical environment, the hopelessness of rural lives, and, especially in the early twentieth century—after the arrival of Freud on the American scene—the repressiveness of so-called Puritan culture.

5 For a brief account of this cultural shift, see Dona Brown and Stephen Nissenbaum, "Changing New England," in William H. Truettner and Roger Stein, *Picturing Old New England: Image and Memory* (New Haven, CT: Yale University Press, 1999). The catalog explores the visual imagery associated with these changes.

6 At first, historians viewed the problem through the eyes of nineteenth-century analysts. Harold Fisher Wilson's 1936 *Hill Country of Northern New England* was the first to offer a detailed account of what he depicted as rural northern New England's economic "winter." Lewis Stilwell's 1948 *Migration from Vermont* pointed to an even earlier "winter" of pre-Civil War outmigration. A later generation of historians, approaching things from the standpoint of the "new social history," discovered new interpretations. Paul Munyon argued in 1978 that there was no significant outmigration from late nineteenth-century New Hampshire, except from the smallest towns. (*A Reassessment of New England Agriculture in the Last Thirty Years of the Nineteenth Century: New Hampshire, A Case Study*, 1978.) Hal Barron's breakthrough *Those Who Stayed Behind* (1984) changed the terms of current understanding, convincingly demonstrating that the cause of population loss in his prototype town of Chelsea was not farming's decline, but a type of deindustrialization: small town industries became casualties of the increasing centralization of production. On a related question, Michael Bell found that the evidence from the census and landscape history of New England did not support the idea of a nineteenth-century decline in agricultural productivity. Michael M. Bell, "Did New England Go Downhill?" *Geographical Review* vol. 79, no 4 (October 1989): 450–466. A later reassessment of outmigration studies using more complex statistical analysis modified Barron's findings somewhat, suggesting multiple interpretations unique to each town. Jeremy Flaherty, "A Multivariate Look at Migration from Vermont," *Vermont History* vol. 74 (Summer/Fall 2006): 127–55.

7 E. R. Pember, "Our Hill Farms," *Eighth Vermont Agricultural Report* (State Board of Agriculture: Montpelier, VT: Watchman and Journal Press, 1884), 362.

8 Thomas H. Hoskins, "Vermont as an Agricultural State," *First Report of the Vermont Board of Agriculture for 1872* (Montpelier, VT: J. and J. Poland, 1872), 571–2.
9 Henry M. Seely, "The Analysis of Fertilizers," *Third Report of the Vermont Board of Agriculture, 1875–1876* (Rutland, VT: Tuttle, 1876), 516. One apparent exception came from a speaker at the Townshend meeting of the Board in 1878, who focused on the loss of population in the area around Jamaica. J. H. Putnam asserted that the soil was indeed worn out by the poor agricultural practices of a previous generation, but that it could be restored by greater attention to grazing land: "upon these old hills make grass a specialty." *Fifth Report of the Vermont Board of Agriculture* (Montpelier, VT: J. and J. M. Poland, 1878), 137.
10 Henry M. Seely, *Fourth Report of the Vermont Board of Agriculture* (Montpelier, VT: 1877), 6.
11 *The Vermont Farmer*, December 15, 1876.
12 Z. E. Jameson, "Vermont as a Home," *First Report*, 556–7.
13 Colonel John B. Mead, "Opportunities for Young Farmers," *Fourth Annual Report*, 521.
14 Henry M. Seely, comment on E. R. Towle, "How to Make Farm Life Pleasant," *First Report*, 551.
15 T. H. Hoskins, comment on Z. E. Jameson, "Vermont as a Home," *First Report*, 564–5.
16 Jameson, "Vermont as a Home," *First Report*, 553.
17 A. William Hoglund, "Abandoned Farms and the 'New Agriculture' in New York State at the Beginning of the Twentieth Century," *New York History* vol. 34, no. 2 (April 1953): 189–90.
18 David Danbom has argued that the Country Life movement included several competing and incompatible ideological positions, and historians have generally followed that interpretative framework. Danbom described two types of urban Country Life reformers. One "hated and feared the cities,"—romantic agrarians who hoped rural values would protect the nation against the corrosive effect of the immigrant-filled industrial cities. The other "hated and feared the countryside,"—progressive social scientists who found rural life insufficiently modern and organized. Allied with them was a third group of businessmen and agricultural scientists who believed that rural problems stemmed not from social inadequacies but from outmoded, inefficient ways of doing business. *The Resisted Revolution: Urban America and the Industrialization of Agriculture, 1900–1930* (Ames: Iowa State University, 1979), 24–42. For a close analysis of the background of the major Country Life reformers, see William L. Bowers, *Country Life Movement in America, 1900–1920* (1974).
19 Sara Gregg presents a comprehensive account of the work of this cohort of agricultural scientists in chapter 3, "Academics and Partisans," in *Managing the Mountains: Land Use Planning, the New Deal, and the Creation of a Federal Landscape in Appalachia* (New Haven, CT: Yale University Press, 2010).
20 Eugene Davenport, "Agricultural Leadership," *14th Biennial Report of the Kansas State Board of Agriculture* (Topeka: Kansas Dept of Agriculture, 1905), 5. As quoted in Danbom, *Resisted Revolution*, 40. Danbom comments that agricultural specialists were not typically so candid about the implications of their positions.
21 Warren played a key role in developing a method that allowed scientists to combine results gathered by numerous workers, often students. As Warren's biographer

describes, the "method was developed to allow students ... to gain the confidence of the people they interview." Bernard F. Stanton, *George F. Warren: Farm Economist* (Ithaca, NY: Cornell University, 2007), 129–30. The survey, with a full discussion of the method, was published as "An Agricultural Survey, Townships of Ithaca, Dryden, Danby, and Lansing, Tompkins County," *Cornell University Agricultural Experiment Station Bulletin* 295 (Ithaca, NY: Cornell University, March 1911).

22 Both changes were the products of legislative action, but the two governors who appointed the Board and the Commissioner also played a role. The first was Edward Curtis Smith, president of the Central Vermont Railroad, and the second was Fletcher Proctor, president of the Vermont Marble Company. Both were sons of powerful manufacturing families, but Proctor supported progressive forestry policies.

23 Historian Paul Searls offers a compelling analysis of this type of "downhill" local elite and their role in Vermont's responses to modernization in *Two Vermonts: Geography and Identity, 1865–1910* (Lebanon, NH: University of New Hampshire Press, 2006).

24 *Randolph Herald and News*, July 17, 1930.

25 Sara Gregg argues in *Managing the Mountains* (2010) that Vermonters ultimately rejected New Deal programs for removing and re-locating farmers on "submarginal" land precisely because Vermont farmers had more local and statewide political influence than did the farmers of the Blue Ridge Mountains. She makes an excellent case for that comparison, but here I am suggesting that there was a decline in that influence over time, even in relatively democratic Vermont.

26 *Agriculture of Vermont: First Annual Report of the Commissioner of Agriculture of the State of Vermont* (Montpelier, VT: Capital City Press, 1909), 5.

27 *Agriculture of Vermont: Fifth Annual Report of the Commissioner of Agriculture of the State of Vermont* (St. Albans, VT: St. Albans Messenger Co., 1913), 3.

28 Vermont State Horticultural Society, *Report of First Annual Meeting* (Montpelier, VT: 1896).

29 In 1909, the legislature allotted $500 each to the Vermont State Horticultural Society and the Vermont Maple Sugar Makers Association, and $1000 to the Vermont Dairymen's Association (by far the most important of these organizations), "for the purpose of promoting, developing and advertising" these agricultural specialties. *First Annual Report of the Commissioner*, 15–17.

30 John Allen Hitchcock, "A Study of Vermont Dairy Farming," *Vermont Agricultural Experiment Station Bulletin 250* (Burlington: Vermont State Agricultural Experiment Station, 1925), 3.

31 *First Annual Report of the Commissioner*, 9.

32 Lewis Cecil Gray, *Introduction to Agricultural Economics* (New York: Macmillan, 1924) 12–14. Gray was a PhD from the University of Wisconsin's groundbreaking agricultural economics program, where he worked with Henry C. Taylor, J. R. Commons and R. T. Ely. Richard S. Kirkendall, "L. C. Gray and the Supply of Agricultural Land," *Agricultural History* vol. 37, no. 4 (October 1963): 206–14.

33 *Thirteenth Biennial Report of the Commissioner of Agriculture of the State of Vermont, 1924–1926* (St. Albans, VT: St. Albans Messenger Co., 1927), 5–6.

34 J. A. Kerr, "in Charge," and Grove B. Jones, *Soil Survey of Windsor County, Vermont* (Washington, DC: Government Printing Office, 1918), 14.

35 David Rice Gardner describes the process by which the Bureau of Soils developed their system in "The National Cooperative Soil Survey of the United States," doctoral thesis,

Harvard University Graduate School for Public Administration, 1957. Reprinted by the Natural Resources Conservation Service, 1998, 40–76. https://www.nrcs.usda.gov/resources/data-and-reports/the-national-cooperative-soil-survey-of-the-united-states.

36 Gardner, "National Cooperative Soil Survey," 72–75.
37 Like many other soils in Vermont, it was made up of glacial till, most often on bedrock of schist in Windsor County. Kerr and Jones, *Soil Survey of Windsor County*, 12–13.
38 The machines specifically mentioned in this report were horse-drawn plows, disc harrows, and mowing machines. Since the land appeared to be most suited to pasture, the soil scientists presumably had in mind the limits of mowing machines in fields that were too steep or stony.
39 William James Latimer was a Clemson University graduate who worked for the U.S. Bureau of Chemistry and Soils; Samuel Oscar Perkins was a graduate of North Carolina College of Agriculture and Mechanical Arts; Kenneth V. Goodman was a graduate of Pennsylvania State College, as was Faye Lesh; Leslie Rockwell Smith graduated from Massachusetts Agricultural College and was a chemist working for the Department of Agriculture.
40 That is how the passage read in the Vermont Commission on Country Life's final report, *Rural Vermont: A Program for the Future* (Burlington: Vermont Commission on Country Life, 1931), 46. As it appeared in the official report of the surveyors, the passage was punctuated differently, giving it a slightly more negative spin: "The soils of Vermont are *best suited to grass, although they* are not inherently as fertile as the Prairie Region." W. J. Latimer et al., *Soil Survey (Reconnaissance) of Vermont* (Washington, DC: U.S. Government Printing Office, 1937), 78.
41 In fact, the 1916 report on Windsor County had reaffirmed to the same judgment: "When properly cared for," the team noted, "these pastures produce well, the grasses growing more continuously through the summer than in warmer climates." Kerr and Jones, *Soil Survey of Windsor County*, 14–15.
42 Latimer, *Soil Survey*, 7.
43 Jamaica soils included Becket, Hermon, and Woodbridge soils, glacial till on gneiss, granite, and schist (not unlike those found in the Harvard Forest in Petersham). *Soil Survey*, 20–23.
44 L. C. Gray, *Introduction to Agricultural Economics*, 242.
45 Sara Gregg points out how frequently the term "submarginal" was used to describe market disadvantages, rather than problems with the soil itself. Gregg, *Managing the Mountains*, 92–93.
46 Samuel A. McReynolds, "Eugenics and Rural Development: The Vermont Commission on Country Life's Program for the Future," *Agricultural History* vol. 71, no. 3 (Summer, 1997): 301 and note 3.
47 A few years after the Commission was formed, aware of the absence of voices from the hill towns, the Commission's executive committee belatedly added a few additional members, including one from Jamaica, Harry Sherwin, the town clerk at the time. Vermont Commission on Country Life, *News Letter* (January 1931). Silver Special Collections Library, University of Vermont.
48 "Meeting of the Vermont Commission on Country Life, House, October 9, 1929." Minutes published as part of the VCCL's *News Letter* (December 1929).
49 "Conserving Vermont," *Burlington Free Press*, October 11, 1929.

50 "Meeting of the VCCL," 20.
51 Harry Pestana Young, "Studies in Vermont Dairy Farming III: Randolph-Royalton Area," *Vermont Agricultural Experiment Station Bulletin 268* (Burlington: Vermont Agricultural Experiment Station, 1927), 3, footnote 1.
52 "Meeting of VCCL," 32–36.
53 Some of the records of the Vermont Commission on Country Life are held in the Vermont State Archives as part of the Eugenics Survey materials. The VSARA holds the draft minutes taken by the secretary, Anna Rome, and the revisions made by committee members to her original minutes. VSARA PRA-005 Eugenics Survey of Vermont and the Vermont Commission on Country Life Records, 1925–1956. Vermont Commission on Country Life Papers, Container PRA-00023, File 8.
54 "Meeting of VCCL," 37.
55 Among the committee reports that afternoon were those from Religious Forces, Fish and Game, Rural-Urban Relations, and Living Standards. Perhaps none of their proposals was particularly controversial. But the other two were the Citizenship Committee, which described its plan to place limits on voting rights; and the Committee on the Human Factor, which featured Henry Perkins describing the progress eugenics research had been making in the state. Neither of those two provoked any comments, either, although the chair seemed to prompt the audience to ask questions of Perkins.
56 Peach created the English program at the military academy and worked there for thirty-seven years before retiring to lead the Vermont Historical Society.
57 "Meeting of VCCL," 60.
58 "Meeting of VCCL," 60.
59 Lemuel James Peet, "Problems of Land Utilization in the Hill Towns of Vermont, Based on a Study of Thirteen Towns in 1929," (master's thesis, University of Vermont, 1930), 3. A shorter revised version was published as C. F. Clayton and L. J. Peet, "Land Utilization as a Basis of Rural Economic Organization," *Vermont Agricultural Experiment Station Bulletin 357* (Burlington: Vermont Agricultural Experiment Station, 1933). The thirteen towns were Fayston, Warren, Roxbury, Granville, Ripton, Goshen, Stockbridge, Sherburne, Shrewsbury, Plymouth, Mt. Holly, Pittsfield, and Wardsboro.
60 Peet, "Problems of Land Utilization," 4–10.
61 Vermont Commission on Country Life, *News Letter* (July 1930).
62 Peet, "Problems of Land Utilization," 28.
63 Peet, "Problems of Land Utilization," 73.
64 Peet, "Problems of Land Utilization," 65.
65 Clayton and Peet, "Land Utilization," 499–50.
66 Peet, "Problems of Land Utilization," 29.
67 Clayton and Peet, "Land Utilization," 138–140.
68 *Rural Vermont*, 148.
69 In "Eugenics and Rural Development: The Vermont Commission on Country Life's Program for the Future," Samuel McReynolds explores the dual influence of eugenics and rural reform on the Commission. I see that influence as triple: eugenics, country life reform, and the new agricultural sciences. *Agricultural History* vol. 71, no. 3 (Summer, 1997).

70 Letter, H. F. Perkins to Eugenics Survey Advisory Committee, enclosed proposal for Comprehensive Rural Survey, May 23, 1927. The Eugenics Survey of Vermont records are located in the Vermont State Archives and Records Administration, but the website compiled by historian Nancy Gallagher, although quite dated in its format, makes select documents more easily available digitally. Nancy L. Gallagher, "Vermont Eugenics: A Documentary History" (2001), www.uvm.edu/~eugenics/insidef.html.

Chapter 3

1 "Jamaica Local Eugenics Survey Completed," *Brattleboro Daily Reformer*, December 26, 1930.
2 Henry C. Taylor, foreword, "Elin Anderson as I Knew Her," in Elin Lilja Anderson, *Rural Health and Social Policy* (Washington, DC: no publisher, 1951).
3 Choate was described in the class notes of the class of 1920 in the *Smith College Alumnae Bulletin* as having left Smith and graduated from Syracuse University, working for Macy's "in the psychological branch of the employment department." *Smith Alumnae Quarterly* vol. 12, no. 1 (November 1920): 104. Her career was centered mostly in New York City, where she appeared in the 1930 census working as a "research economist" for the United States government; in the 1940 census, when she was listed as "interviewer;" and in 1950, when she worked as "supervisor of state employment services."
4 "Letter, H. F. Perkins to the Advisory Board of the Eugenics Survey, January 4, 1929." Nancy L. Gallagher, "Vermont Eugenics: A Documentary History" (2001), www.uvm.edu/~eugenics/insidef.html.
5 For the most comprehensive account of the story of Henry Perkins and the Eugenics Survey of Vermont, see Nancy Gallagher, *Breeding Better Vermonters: The Eugenics Project in the Green Mountain State* (Lebanon, NH: University Press of New England, 1999). Samuel A. McReynolds lays out the relationship between the eugenics movement and the Vermont Commission on Country Life: "Eugenics and Rural Development: The Vermont Commission on Country Life's Program for the Future," *Agricultural History* vol. 71, no. 3 (Summer, 1997). An important newer study of eugenics in the larger context of state politics and public policy is Mercedes De Guardiola, "*Vermont for the Vermonters:*" *The History of Eugenics in the Green Mountain State* (Barre: Vermont Historical Society, 2023).
6 "Letter, H. F. Perkins to the Eugenics Survey Advisory Committee (Dr. E.A. Stanley), February 12, 1929." Gallagher, "Vermont Eugenics," www.uvm.edu/~eugenics/insidef.html.
7 "Letter, H. F. Perkins to Eugenics Survey Advisory Committee, May 23, 1927." Gallagher, "Vermont Eugenics," www.uvm.edu/~eugenics/insidef.html.
8 Vermont Commission on Country Life, *Rural Vermont: A Program for the Future* (Burlington: Vermont Commission on Country Life, 1931), 10.
9 *Rural Vermont*, 27.
10 Perkins explained the politics of the situation in a letter to the Eugenics Survey's advisory board: "The Vermont Commission is in the position of needing active

support from every quarter." The Eugenics Survey could not allow its support for a new sterilization bill to "arouse suspicion" of the Country Life Commission "among any important group, economic, social, political or religious," referring chiefly to the Catholic church, which maintained the most sustained opposition to sterilization. "To the Advisory Board of the Eugenics Survey, January 4, 1929." Gallagher, "Vermont Eugenics," www.uvm.edu/~eugenics/primarydocs/olhfpac010429.xml.

11 In a fascinating analysis of the eugenics family studies, Nicole Hahn Rafter points out the contradiction between the tendency of early twentieth-century progressives to romanticize rural life and their distaste for poor rural people. Nicole Hahn Rafter, *White Trash: The Eugenic Family Studies, 1877–1919* (Boston: Northeastern University Press, 1988).

12 Perkins cited the supposedly poor medical records of Vermont's rural, largely "Yankee" draftees during World War I as the catalyst for his interest in eugenics studies. Gallagher, *Breeding Better Vermonters*, 39–40.

13 In Jamaica, a total of 24 out of 158 interviews included at least one reference to non-Yankee descent, including Dutch (1), Irish (1), Polish (1), Scottish (1), Swedish (1), English Canadian (2), French Canadian (5), Finnish (3), German (5), and "Indian" (4) descent.

14 If Dempsey had not spoken up, others probably would have done so. Perkins was aware of a general resistance to his diagnosis of rural decline. "First meeting of the Executive Committee of the Vermont Commission on Country Life, May 18, 1928," VSARA, PRA-005, Container PRA-00023, file 10.

15 Sara Gregg argues, for example, that Vermonters ultimately rejected New Deal programs for removing and re-locating farmers on "submarginal" land precisely because Vermont farmers had more local and statewide political influence than did the farmers of the Blue Ridge Mountains. Sara Gregg, *Managing the Mountains: Land Use Planning, the New Deal, and the Creation of a Federal Landscape in Appalachia* (New Haven, CT: Yale University Press, 2010).

16 Minutes of meeting of Land Utilization committee held in Burlington, Vermont, June 27, 1930. VSARA, PRA-005, Container PRA-00029, file 26.

17 Henry F. Perkins, "Contributory Factors in Eugenics in a Rural State: A Decade of Progress in Eugenics: Scientific Papers of the Third International Congress of Eugenics," August 22, 1932, *A Decade of Progress in Eugenics: Scientific Papers of the Third International Congress of Eugenics*, (New York, 1934), 186–187.

18 Elin Anderson, *We Americans* (Cambridge: Harvard University Press, 1937). A glowing review by Robert P. Tristram Coffin mentions that it won the John Anisfield Prize "for the best book in the field of race relations in America." *Atlantic Monthly*, April 1938.

19 Martha M. Wadman, "Towns Suggested for Study: Field Notes on Waitsfield, Jamaica, and Cornwall," 1928. Gallagher, "Vermont Eugenics," https://www.uvm.edu/~eugenics/primarydocs/ofmswcj000028.xml.

20 Genieve Lamson, *A Study of Agricultural Populations in Several Vermont Towns* (Burlington, VT, 1931). Lamson, like Anderson, took her study in a direction different from what Perkins had in mind.

21 Crockett was a member of the Committee on Summer Residents, the Land Utilization Committee, and the Committee on Traditions and Ideals. Jones was a member of both

the Land Utilization Committee and the Farm Production Committee; Ingalls was a member of the Recreation Committee and an education subcommittee; and Bradlee was a member of the Executive Committee and the Farm Production Committee. Frazier was a member of both the Committee on Religious Forces and the Committee on Community Trusts.

22 Sara Gregg presents the Vermonters on the Country Life Commission as "locals, members of the thinking classes." In my view, she somewhat overestimates the degree to which Vermont elites identified with and represented the interests of hill farmers. Gregg, *Managing the Mountains*, 41. The Commission added Harry Sherwin, the town clerk of Jamaica at the time. Vermont Commission on Country Life, *News Letter* (January 1931).

23 "Selective Migration from Three Rural Vermont Towns and its Significance," *Fifth Annual Report of the Eugenics Survey of Vermont* (Burlington: University of Vermont, 1931), 13.

24 John Hooper, "County Correspondents Report Doings in Rural America," *Brattleboro Daily Reformer*, October 12, 1938, 4. I am grateful to Karen Ameden for showing me this story.

25 Anderson's account is strikingly similar to that of a team of researchers in the Blue Ridge Mountains at the same time, not eugenicists but "two psychologists, a nutritionist, a psychiatrist and sociologists." They wrote: "To the visitor the deep silence and the drowsiness of the mountains are fascinating. After a few days the hollow appears an ideal bed chamber designed by nature itself—with a drowsy blue haze on the mountain side and the sweep of the wind to lull one to sleep." Mandel Sherman and Thomas R. Henry, *Hollow Folk* (New York: Thomas Crowell and Co., 1933), 9.

26 "Selective Migration," 74.

27 "Meeting of the Vermont Commission on Country Life, October 9, 1929." VCCL *News Letter* (December 1929), 32–36.

28 "Meeting of the VCCL," 23.

29 "Selective Migration," 78.

30 "Selective Migration," 78.

31 The version that became the annual report of the eugenics survey seems to have been written mostly by Anderson, but the version that was integrated into the chapter called "The People" in *Rural Vermont* may have been influenced by Perkins or perhaps by Paul Moody, president of Middlebury College and chair of the Committee on the Human Factor. There was also an additional editor named John Holden, who "wove together" the essay, as Perkins reported. Henry Perkins, "Foreword," in "Selective Migration," iii.

32 *Rural Vermont*, 29–30.

33 Lemuel J. Peet, "Problems of Land Utilization in the Hill Towns of Vermont" (master's thesis, University of Vermont, 1929), 65.

34 Later, in retreat from some of the harsher implications of eugenics, Perkins came up with a slightly more humane-sounding phrasing of this diagnosis: "A fine old pioneer stock deserves an environment commensurate with its quality. . . . If their home surroundings are poor and life nothing better than a perpetual fight for the merest necessities, their native ambition may be so dampened as to make them indifferent to their future and that of their children." Perkins, "Contributory Factors," 187.

35 It is not clear who designed the questionnaire.

36 Anderson's career was dedicated to working on rural social problems. After her work with the Eugenics Survey, she moved to the USDA's Extension Service in Washington, DC. She later piloted rural health projects with the Farm Foundation in Chicago and Nebraska. Henry C. Taylor, foreword, "Elin Anderson as I Knew Her," in Elin Lilja Anderson, *Rural Health and Social Policy*, (Washington, DC: no publisher, 1951).

37 Obituary of Elliott N. Choate, *Fitchburg Daily Sentinel*, December 20, 1919. He died when Marjorie Choate was 21 years old.

38 As Nicole Hahn Rafter suggests, these jobs were perceived as "natural" to women: women were thought to be better at extracting information from people being investigated; they were considered more detail oriented; and the positions were often construed as "assistants" to male professionals. That was certainly true in Perkins' case: he hired only women in the Eugenics Survey, but they were all placed in the position of being his "assistants." Rafter, *White Trash*, 21.

39 "Selective Migration," 69.

40 Interview #118. Unless otherwise noted, all numbered interviews are from the Eugenics Survey of Vermont Migration Study field notes for Jamaica. VSARA; PRA-005: Container 00015, files 3 and 4. Interviews from Cornwall are also located in Container PRA-00015, and those from Waitsfield are located in Container PRA-00014. Note also that in some cases, I have assigned a pseudonym to the person who was interviewed.

41 Interview #169A.

42 Interview #197.

43 Interview #196.

44 "Local Eugenics Survey Completed," *Brattleboro Daily Reformer*, December 26, 1930.

45 Interview #223.

46 Interview #223.

47 36 percent, or 57 out of 158 interviews. This includes eleven comments from residents and forty-six from the interviewers. It appears that the interviewers were not raising these questions openly with the residents, but rather recording their private judgments in the field notes.

48 Twenty-nine were recorded by Anderson, fourteen by Choate, and fourteen by Rome.

49 John B. Mead, "Opportunities for Young Farmers," in Henry M. Seely, *Report of the Vermont State Board of Agriculture, 1875–1876* (Rutland, VT: Tuttle Co., 1876), 516.

50 Interview #99.

51 Interview #198.

52 Interview #94.

53 Interview #43.

54 Interview #220.

55 "Selective Migration," 23.

56 Interview #13 Waitsfield. VSARA; PRA-005; Container PRA-00014, file 34.

57 Some households gave an answer for more than one individual, some individuals gave more than one answer, and it is often difficult to tell which answer belongs to whom. For some reason, the interviewers recorded significantly fewer responses to this question in Cornwall.

58 VSARA PRA-005: Container PRA-00015, File 27.

59 I was unable to determine what criteria the team used to distinguish migrants from people who had always lived there (whether, for example, they counted someone who was born in an adjoining town as an "immigrant," or how long someone had to live in town to be counted as someone who "stayed in town" rather than "came to town.") I did not attempt to analyze the emigrant files in detail because they are quite incomplete, and mostly based on relatives' reports. By the team's count, 58 percent of the people who left town cited economic reasons for doing so, and 47 percent of those who migrated into town gave an economic reason, while "the group who remained" gave economic reasons for staying in town only 18 percent of the time. "Selective Migration," 57–59. By my count, 31 percent of all the residents who answered the question—eighty-six individual responses—mentioned some kind of economic reason.

60 Nine residents of Cornwall and four Waitsfieldians said they had never considered quitting the farm. No residents of Jamaica responded exactly that way; nine did say they had never considered leaving *town*. Of the three towns, Cornwall was the most devoted to farming; perhaps they did not perceive a difference between leaving the farm and leaving town.

61 "Selective Migration," 23–24.

62 Interview #192 Waitsfield.

63 The numbers for Jamaica were: seven preferred the region; eleven preferred country life; eighteen "liked farming." The numbers for Waitsfield and Cornwall were rather similar, but much larger majorities in those towns specified that they liked farming, while in Jamaica there were many more residents who preferred the region or "country life."

64 Interview #127.

65 Interview #24.

66 Interview #7.

67 Interview #7.

68 Interview #108.

69 Interview #121.

70 Interview #53.

71 Interview #8.

72 Interviews #43 and #220.

73 Interview #146.

74 She might have been mistaken about the "Christian Science woman." Jamaica was home to a long-established community of Seventh Day Adventists. They often worked for short or longer periods of time at the Adventist hospital near Boston, and one of them might well have given this advice.

75 It is not clear how the eugenics team defined "native-born," but I counted as native-born the heads of household who were born in the adjoining towns of Wardsboro, Townshend, Windham, Londonderry, Stratton, and Winhall (which includes the village of Bondville).

76 Interview #58.

77 Interview #55.

78 Interview #4.

79 Interview #6.
80 Interview #183.
81 Interview #124.
82 Interview #127.
83 Interview #185.
84 Interview #99.
85 Interview #117.
86 Interview #44.

Chapter 4

1 Interview #110. Unless otherwise noted, all numbered interviews are from the Eugenics Survey of Vermont Migration Study field notes for Jamaica. VSARA; PRA-005; Container PRA-00015. I used a pseudonym here because the son was interviewed by the Eugenics Survey. The family name is recorded in Interview #110, and that can be used to trace the father's probate and census records.
2 Hamilton Child, *Gazetteer and Business Directory of Windham County, Vermont* (Syracuse, NY: 1884), 421–433.
3 *Walton's Vermont Register, Business Directory, Almanac, and State Year-book for Farmers, Business and Professional Men* (Rutland, VT: Tuttle Co, 1885), 94. The 1930 number comes from the Eugenics Survey field notes and excludes four men who were operating small semi-portable sawmills or other small machines requiring just one worker to operate.
4 In comparison, Hal Barron's Chelsea lost almost no farms between 1850 and 1900. Barron, *Those Who Stayed Behind: Rural Society in Nineteenth-Century New England* (Cambridge: Cambridge University Press, 1984), 57.
5 In the years when the agricultural census was not available, I counted the number of farms in the population censuses.
6 63 percent of Jamaica farmland was in woods in 1930, including pastured woods, and 37 percent in some kind of open land—tilled, idle, pastured, or grazed. 102 farms were counted, with a total of 15,952 acres in farmland—156 acres per farm on average. *Fifteenth Census of the United States: 1930*, Agriculture, Vol. 1 (Washington, DC: Bureau of the Census, 1931), 642.
7 Probate Records, Windham County, Vermont; February 12, 1891; 249–50; Folder 13, Bundle 10, 35/237–35/332. I have not included the family name in this citation to protect the name of this man's son, who was interviewed by the eugenicists.
8 C. F. Clayton, "Farm Abandonment Goes by Definite Stages in Vermont's Hill Towns," *Yearbook of Agriculture, 1931* (Washington, DC: United States Government Printing Office, 1931), 225–7.
9 The acreage of the home farm was reported differently at different times. In the 1870 agricultural census it was reported as eighty acres, with sixteen acres wooded; in the 1880 agricultural census as just sixty-one acres, with eight acres in woods; and the 1890 probate record reported the farm as eighty-five acres.
10 W. J. Latimer et al., *Soil Survey (Reconnaissance) of Vermont* (Washington, DC: Government Printing Office, 1937), 66. The thirteen-town study found that well

over half of the land used for grazing in each of the thirteen towns in 1929 was wooded, and in some towns much more. Woodland grazing ranged from 57 percent of all grazing land in the town of Warren to 85 percent in Sherburne. Lemuel Peet, "Problems of Land Utilization in the Hill Towns of Vermont" (master's thesis, University of Vermont, 1929), 115. Even in Vermont as a whole, 47 percent of all land used as pasture was wooded.

11 Interview #46.
12 The interviewers rarely followed this line of inquiry in the Cornwall or Waitsfield interviews, although the "condition of the farm" section is on the questionnaire forms.
13 Interviews #3; #12; #22; #95; #137.
14 Interview #97.
15 Peet, "Problems of Land Utilization," 107. As Peet commented, studies in other places had showed a close relationship between quality of land and income, but here that was not the case.
16 J. A. Kerr, Grove B. Jones, *Soil Survey of Windsor County, Vermont* (Washington, DC: Government Printing Office, 1918), 15.
17 One article in a Vermont Commission on Country Life newsletter did report that finding: "Contrary to general opinion it was not found that hill farms had been deserted because of loss of fertility of the soil." Quoting from a six-page abstract of the thirteen-town study, the author explained that "stony fields" were an important problem, but so was the "inability to get sufficient business." Vermont Commission on Country Life, *News Letter* (July 1930), 5.
18 Interview #118.
19 Interview #12. Hill farmers used artificial fertilizer mostly on corn and potatoes. *Soil Survey of Windsor County*, 11.
20 Interview #104.
21 Interview #4.
22 Interview #43.
23 Interview #22.
24 The 1930 agricultural census introduced these new categories, to be determined by 50 percent of the value of the farm's production. On the population census schedules, however, "general farm" was the default term used for nearly all farms, at least in the Vermont towns I investigated.
25 Richard M. Judd, *The New Deal in Vermont: Its Impact and Aftermath* (Garland: New York and London, 1979), 7.
26 Very slightly more than in the other two towns studied by the eugenicists: 80 percent in Jamaica; 79 percent in Waitsfield; 72 percent in Cornwall.
27 Ten farmers were recorded by the eugenicists as selling cream; eight selling butter; and eighteen selling fluid milk. The interviewers did not always record what kind of dairy products the farmer was selling, but these numbers are probably representative.
28 Interviews #47 and 48.
29 Interview #43.
30 Interview #118.
31 Interview #123. William L. Bowers, *Country Life Movement in America, 1900–1920* (Kennikat Press, 1974), 14.
32 Interview #122.

33 Interview #118.
34 Interview #80.
35 Interview #22.
36 Two out of every five farms they studied had small mortgages, but only 3 farms out of the 146 were mortgaged up to the value of the farm. Peet, "Problems of Land Utilization," 136, 138.
37 Peet, "Problems of Land Utilization," 126. In that study, only households that grossed above $3000 per year showed a net increase at the end of the year. In Jamaica, there were no farmers who grossed above $3000 in 1930.
38 Peet, "Problems of Land Utilization," 138–9.
39 Interview #95.
40 The interviewers did sometimes speculate that a resident was not capable of keeping track of the numbers, but they mostly referred to people who had migrated from Quebec and who were not fluent in English. In the Jamaica interviews, nobody was labeled illiterate, but two men were described as barely literate. One "could read but little," but "has an excellent memory." The other—a first-generation migrant from Quebec—"cannot write and reads with difficulty." Interviews #160 and #83. The cumulative chart in the eugenics survey shows six individuals in Jamaica as illiterate— three "residents" and three "immigrants." "Selective Migration from Three Rural Vermont Towns," *Fifth Annual Report of the Eugenics Survey of Vermont* (Burlington: University of Vermont, 1931), 46.
41 "Selective Migration," 46. A quarter of residents had attended some high school.
42 For example, one farmer told Marjorie Choate that he was experimenting with sending half of his milk to Londonderry and half to Brattleboro to see which one would pay better. Several reported that they were experimenting with marketing milk only in winter to take advantage of better prices. (Most farmers were still milking only in the summer season when feed was most abundant.) One was planning to grow beans instead of potatoes since the market for potatoes had been weak the previous year. Interviews #122, #190, #45.
43 Interview #122.
44 The author twice cited the Biblical commandment from Exodus: "Thou shalt not muzzle the ox that treadeth out the corn," interpreting it to mean that the laborer is due a fair share in the harvest. "Value of Farm Products in 1899," *Twelfth Census of the United States: Agriculture*, Vol. V (Washington, U.S. Census Office, 1902), cxx.
45 I counted households where the main income was from farming, and households with both farming and wage income. I did not include households where the income came from a business or off-farm wages only.
46 Historian Christopher Harris points out that off-farm work was always essential to the mixed-farming systems that had evolved in New England over the previous three centuries, and indeed that off-farm income has been essential to smallholders in all parts of the globe, helping to keep their households afloat and able to respond to a host of challenges. Christopher Harris, "The Road Less Traveled By: Rural Northern New England in Global Perspective" (PhD dissertation, Northeastern University, 2007) 126, 237.
47 Interview #119.
48 Interview #80.

49 Interview #135.
50 The 1884 Child's *Gazetteer* listed twelve men in Jamaica who did enough lumbering work to include as an occupation alongside farming.
51 "Fern Picking Bonanza in Some Rural County Towns," *Brattleboro Daily Reformer*, September 17, 1923, 5.
52 *Brattleboro Daily Reformer*, September 17, 1923, 5.
53 Interview #194.
54 Interview #99.
55 "Selective Migration," 70.
56 An outstanding historical account of the impact of the 1927 flood on Vermont is Deborah Clifford and Nicholas Clifford, *"The Trouble Roar of the Waters:" Vermont in Flood and Recovery, 1927–1931* (Lebanon, NH: University Press of New England: 2007).
57 For examples, Interview #118, #130, #169A.
58 Interview #223.
59 Peet, "Problems of Land Utilization," 193.
60 *Brattleboro Daily Reformer*, November 30, 1963. This is a pseudonym, so to locate the obituary, it will be necessary to locate the family name in the Jamaica records: Interview #22.
61 Interview #20.
62 E. R. Pember, "Our Hill Farms," *Eighth Vermont Agricultural Report* (Montpelier, VT: Watchman and Journal Press, 1884), 366.
63 Interview #95.
64 The team reported that thirty-three of their forty-three households were modern farms for whom dairy was "practically the only source of income." Marianne Muse, Isabelle Gillum, "Food Consumption of 50 Vermont Farm Households," *Vermont Agricultural Experiment Station Bulletin 327* (Burlington: Vermont Agricultural Experiment Station, 1931), 6.
65 Muse, Gillum, "Food Consumption," 11.
66 The Muse-Gillum study found that their mostly prosperous farmers spent an average of $320 on food and consumed $374 worth of food from home—$694 in all. The Peet study confirms a similar finding for the thirteen hill towns, but at a lower level of expenditure, and as the Muse study found, a little less of the food coming from the farm. There, the total was $529, average cash spent on food was $276, and the average value of food furnished by the farm was $253—48 percent provided from the farm. C. F. Clayton and L. J. Peet, "Land Utilization as a Basis of Rural Economic Organization," *Vermont Agricultural Experiment Station Bulletin 357* (Burlington: Vermont Agricultural Experiment Station, 1933), 120.
67 Dorothy Emery, "A Food Budget for Vermont Farm Families," *Vermont Agricultural Experiment Station Bulletin 393* (Burlington: Vermont Agricultural Experiment Station Bulletin, 1935), 6.
68 Interview #12.
69 That data was collected during the Depression, so it may have been an unusually high number, but the eugenics surveys showed similar results in 1930, before the Great Depression had re-shaped many household budgets. Marianne Muse and Margaret E. Openshaw, "Incomes and Expenditures of 299 Vermont Village Families," *Vermont*

Agricultural Experiment Station Bulletin 450 (Burlington: Vermont Agricultural Experiment Station, 1939), 24.
70. Interview #187 Waitsfield; Interview #92 Jamaica; Interview #13 Cornwall.
71. Interview #44.
72. The general pattern is clear, but the details are not exact. The amount of information varied a great deal from one interview to the next. Some of those I counted here specified that they owned over 20 acres but were raising "only" a garden or a garden with cow or chickens. Some reported owning over 20 acres but using 5 or fewer acres in any way, including pasture, grazing, or maple production. It would be more accurate to count somewhere between thirty and forty households living on full-scale farms who were now not "farming," but "gardening" or even "foraging" for a living.
73. Clayton and Peet, "Land Utilization," 118–119.
74. Interview #35.
75. Interview #83.
76. They also counted eight farm laborers working on their home farms (that is, sons), eighteen farm laborers on other farms, and eight retired farmers living in town. These numbers are from the penciled tallies for all three towns. VSARA, PRA-005 Eugenics Survey of Vermont; Migration Study, Container PRA-00014, File 27.
77. I have assigned a pseudonym to the census taker because he was interviewed by the eugenicists. His real name is recorded on the census schedules.
78. *Fifteenth Census of the United States: 1930 Agriculture*, vol. 1 (Washington: U.S. Government Printing Office, 1931) 1.
79. That was not always the case. In other locations, census enumerators sometimes marked the box for "Is this family living on a farm?" with a "yes," but did not give the farm a number for the agricultural census, evidently judging that it did not even meet the very minimal requirements of the census.
80. Interview #119.
81. "1930 census: numerator instructions," Steven Ruggles, Sarah Flood, Matthew Sobek, Daniel Backman, Annie Chen, Grace Cooper, Stephanie Richards, Renae Rodgers, and Megan Schouweiler. IPUMS USA: Version 15.0 [dataset]. Minneapolis, MN: IPUMS, 2024. https://usa.ipums.org/usa/voliii/inst1930.shtml.
82. In his 1930 interview with the eugenicists, the 1920 enumerator told Margaret Choate that he also did not "wish to leave J. His folks have tried to get him to do so but he refuses." Interview #214.
83. Interview #23.

Chapter 5

1. The figure was reported as $567 in C. F. Clayton and L. J. Peet, "Land Utilization as a Basis of Rural Economic Organization," *Vermont Agricultural Experiment Station Bulletin 357* (Burlington: Vermont Agricultural Experiment Station, 1933), 84. Table 34 shows a total net income of $478 from the farm and off-farm work, plus $89 from pensions, boarding, and other sources. In Lemuel Peet's master's thesis, the net farm income average is reported to be $601. There Peet also referred to studies in New England and Vermont in the previous few years which found the average

net income of farm families in the region was $800-$1000. By his count, fewer than a quarter of the 162 hill farmers attained that average income. Lemuel James Peet, "Problems of Land Utilization in the Hill Towns of Vermont, Based on a Study of Thirteen Towns in 1929" (master's thesis, University of Vermont), 1930. 85–86, 140—Table 60.

2 Marianne Muse, "Standard of Living on Specific Owner-Operated Vermont Farms," *Vermont Agricultural Experiment Station Bulletin 340* (Burlington: Vermont Agricultural Experiment Station, 1932), 10. Gross income average was $3,500, minus $2,096 operating expenses.

3 In her published account Anderson asserted that there were no such families in prosperous Cornwall, but the notes for this family contradict her, specifying that this family was being "helped by the town." Interview #167 Cornwall. VSARA; PRA-005; Container PRA-00015.

4 Interview #131A Waitsfield. VSARA; PRA-005; Container PRA-00014.

5 The four people "on the town" included one adult man who was "feebleminded and epileptic;" one woman who had been left paralyzed by a "shock," or stroke; one who was disabled by "rheumatism," perhaps rheumatoid or osteoarthritis; and one man who was "too old to work." The households received $7 per week from the town for three of these boarders, and $15 for the woman who had the stroke, as she was "completely helpless." Unless otherwise noted, all numbered interviews are from the Eugenics Survey of Vermont Migration Study field notes for Jamaica. VSARA; PRA-005; Container PRA-00015, files 3 and 4. Interviews #103, #148, #150.

6 These numbers are adequate for general comparison, but they are not precise. Especially in Jamaica, some incomes were recorded as gross, some as net, some as wages—and some as an unclear combination of those categories. The figures were slightly clearer in the other towns, mostly because more farm households were able to report both gross and net incomes.

7 Comparing figures in either category is problematic because so few Jamaica households used net figures, and so few in the other two towns used gross figures. The five very large Waitsfield farms that reported their incomes as gross figures averaged a whopping $3,832 per farm. That included the property held by the Commissioner of Agriculture, E. H. Jones, who reported a gross figure of $10,000 that year, but was uncertain of his net income since he had plowed so much of the farm's earnings into improvements.

8 Lillian H. Johnson and Marianne Muse, "Cash Contribution to the Family Income Made by Vermont Farm Homemakers," *Vermont Agricultural Experiment Station Bulletin 355* (Burlington: Vermont Agricultural Experiment Station, 1933).

9 Interview #174. Rome also recorded that Mrs. White worked as a domestic nurse (meaning an untrained caretaker) "whenever she gets the chance," but recorded no income for that.

10 Ellis Lore Kirkpatrick, *The Farmer's Standard of Living* (New York and London: Century, 1929), 1–3. In explaining the controversies over how to measure standard of living, Kirkpatrick concluded that the simple numerical cost of living was an adequate measure of all the other matters more difficult to enumerate, but he acknowledged that many other specialists disagreed.

11 Muse, "Standard of Living," 3.

12 Although this study was quite inconclusive, it seemed to have general appeal: "No Best Way Seen for Bedmaking," *Brattleboro Daily Reformer*, June 13, 1949, 8.
13 Muse, "Standard of Living," 14–15.
14 Muse, "Standard of Living," 25.
15 Marianne Muse, "Farm Families of Two Vermont Counties: Their Incomes and Expenditures," *Vermont Agricultural Experiment Station Bulletin 490* (Burlington: Vermont Agricultural Experiment Station, 1942) 2.
16 Kirkpatrick, *Farmer's Standard of Living*, 10, 53.
17 Clayton and Peet, "Land Utilization," 91. The numbers were $567 average net cash income and $482 average non-cash income "furnished by" the farm.
18 Clayton and Peet, "Land Utilization," 120–121.
19 Nutritional values as they were understood at the time, of course. Marianne Muse and Isabelle Gillum, "Food Consumption of Fifty Vermont Households," *Vermont Agricultural Experiment Station Bulletin 327* (Burlington: Vermont Agricultural Experiment Station, 1931); Marianne Muse and Ruth Johnston, "Diets of Vermont Farm Families," *Vermont Agricultural Experiment Station Bulletin 573* (Burlington: Agricultural Experiment Station, University of Vermont, 1953).
20 Muse and Johnston, "Diets of Vermont Farm Families," 11.
21 Muse and Johnston, "Diets of Vermont Farm Families," 25. In a typical June diet "among families with the poorest meals," for example, Muse found they consumed 82 percent fewer fresh vegetables than recommended (it was June, after all, and they didn't like greens), 26 percent more canned fruits and vegetables, 11 percent more potatoes, and 130 percent more sugar than recommended.
22 Muse and Johnston, "Diets of Vermont Farm Families," 30.
23 Muse and Johnston, "Diets of Vermont Farm Families," 30.
24 John A. Hitchcock, "A Study of Vermont Dairy Farming," *Vermont Agricultural Experiment Station Bulletin 250* (Burlington: Vermont Agricultural Experiment Station, 1925), 12. Muse, "Standard of Living," 15.
25 Muse and Gillum, "Food Consumption," 7. That does not imply that they had indoor water toilets, although it might have been possible to rig one up if the water pressure was sufficiently strong.
26 Clayton and Peet, "Land Utilization," 121. Marianne Muse and her colleague Margaret Liston found a few years later that 65 percent of the farm household sample had telephones, and 80 percent had water in the kitchen. But in this group, a much larger percentage had electricity—65 percent—perhaps because they were located near villages. Marianne Muse and Margaret Liston, "The Relative Economy of Household Production and of Purchase of White Bread," *Vermont Agricultural Experiment Station Bulletin 392* (Burlington: Vermont Agricultural Experiment Station Bulletin, 1935), 11.
27 Fourteen households were reported specifically as having electricity, twelve were reported to have radios, and thirteen were specifically reported to have no radio or running water or electricity.
28 Warren E. Booker, *Historical Notes: Jamaica* (Brattleboro, VT: Hildreth and Co., 1940), 77, 79.
29 Interview #161.
30 Interview #132.
31 Muse and Liston, "Relative Economy," 11.

32 Interview #31.
33 Peet, "Problems of Land Utilization," 29.
34 The situation in Hal Barron's Chelsea had been similar a generation earlier: it had been the decline of manufacturing in town that led to population decline. That process was even plainer and more extreme in the hill towns in the 1920s: while Chelsea had declined to half its size before stabilizing in around 1900, Jamaica fell by two-thirds before it stabilized in 1930.
35 Interview #1.
36 Interview #157.
37 Interview #220.
38 "Fern Picking Bonanza," *Brattleboro Daily Reformer*, September 17, 1923, 5.
39 Interview #22.
40 Interview #204.
41 "Selective Migration from Three Rural Vermont Towns," *Fifth Annual Report of the Eugenics Survey of Vermont* (Burlington: University of Vermont, 1931), 46.
42 Clayton and Peet, "Land Utilization," 54.
43 Interview #103.
44 Interviews #153 and #157.
45 Interview #31.
46 The Seventh Day Adventists have a long history in Jamaica, and it is likely that both staff and clients at the Sanatorium were mostly also Adventists. Interview #48.
47 Seven people in Waitsfield and four in Cornwall reported that retirement was their reason for being in town; the number in Jamaica was thirteen.
48 Interview #106.
49 Interview #103.
50 Interview #6 Waitsfield.
51 See, for examples, Interviews #4, #32, #35, #43, #45, #47.
52 Robert Riley, "Kinship Patterns in Londonderry, Vermont, 1772–1900: Changing Family Relationships," PhD dissertation, University of Massachusetts—Amherst, 1980, 168.
53 Interview #24.
54 Interview #88, Jamaica Emigrant Field Notes. VSARA, PRA-005; Container PRA-00015, File 3.
55 The emigrant files include 59 second-hand reports from relatives and friends about people who left town. They mostly record the reason for leaving and the current work or marital status of the emigrant. The numbers were calculated by the eugenics team and are located in handwritten notes: PRA-00014, Files 27 and 33.
56 Clayton and Peet, "Land Utilization," 54.
57 In general, farm wages and factory wages kept pace with each other until the end of World War I. After that, industrial work took off in comparative terms and stayed significantly higher through the Depression and beyond. Thurston Madison Adams, "Prices Paid by Vermont Farmers for Goods and Services and Received by Them for Farm Products, 1790–1940," *Vermont Agricultural Experiment Station Bulletin 507* (Burlington: Vermont Agricultural Experiment Station, 1944), 92–94.
58 One giant federally sponsored 1936 study included both several hundred village households and nearly a thousand farm households in northern Vermont. Overall,

villagers earned (including both cash and non-cash income) an average of $1,617 that Depression year. Villagers at the lower occupational level of "wage earners" earned an average of $1,309. Farm families earned just $1,217 on average. The data was reported in two different bulletins: Marianne Muse and Margaret Openshaw, "Incomes and Expenditures of 299 Vermont Village Families," *Vermont Agricultural Experiment Station Bulletin* 450 (Burlington: Vermont Agricultural Experiment Station, 1939); and Marianne Muse, "Farm Families of Two Vermont Counties, Their Incomes and Expenditures," *Vermont Agricultural Experiment Station Bulletin* 490 (Burlington: Vermont Agricultural Experiment Station, 1942).

59 U.S. Bureau of Labor statistics for 1929 indicate that average full-time hours per week for all employees in machine shops were 50.3 hours per week, and full-time earnings per week averaged $32.09. For "laborers" the figure was about $23.68 per week, around $1184 per year. Average pay for "laborers" in foundries was about $25/week or around $1250 per year. United States Department of Labor, *Wages and Hours of Labor in Foundries and Machine Shops, 1929* (U.S. Government Printing Office: Washington, DC, 1930), 1–2, 3, 6, 60.

60 Interview #45.

61 Interview #97.

62 This was the average annual wage for people who reported wages only—and reported them as an annual figure rather than a day wage only (only twenty workers out of the one hundred twenty-five who offered information about income). This category includes mostly more or less year-round jobs in the sawmill or lumbermill, along with more seasonal but higher-paid work for the telephone company.

63 U.S. Bureau of Labor Statistics, *Wages and Hours of Labor in the Paper and Pulp Industry, 1923* (Washington, DC: Government Printing Office, 1925), 38.

64 Interview #3.

65 Twenty-four in Jamaica as opposed to seven in Waitsfield and eleven in Cornwall.

66 Interviews #97, #125, #45.

67 One interviewee worked on the road in summer and cut "some" cordwood in winter (Interview #52). One cut cordwood in winter and did some other kind of labor at other times of the year (Interview #92). One reported working with his truck in good weather and logging in the winter (Interview #161). 54 men described their positions as "laborer."

68 Richard Judd suggests that for hill farmers, "multiple income possibilities preserved an element of choice important to their self-image." Richard Judd, *Second Nature: An Environmental History of New England* (Amherst and Boston: University of Massachusetts Press, 2014), 139.

69 Interview #220.

70 Clayton and Peet, "Land Utilization," 120–121.

71 Interview #129.

72 Interview #83.

73 The number was seven out of the eighty-eight outmigrants whom the interviewers were able to identify as having "gainful employment." That category excluded all but one woman, so it does not reveal anything about women's reasons for leaving. "Selective Migration," 28.

74 Peet, "Problems of Land Utilization," 69.

75 Clayton and Peet, "Land Utilization," 138.
76 Interview #118.
77 Interview #124.
78 Interview #130.
79 Peet, "Problems of Land Utilization," 23.
80 John A. Hitchcock, "A Study of Vermont Dairy Farming," *Vermont Agricultural Experiment Station Bulletin 405* (Burlington: Vermont Agricultural Experiment Station, 1925), 22.
81 E. W. Bell, "Studies in Vermont Dairy Farming IV: Cabot-Marshfield Area," *Vermont Agricultural Experiment Station Bulletin 283* (Burlington: Vermont Agricultural Experiment Station, 1928).
82 Hitchcock, "A Study of Vermont Dairy Farming," 24.
83 Philip K. Hooker, "Studies in Vermont Dairy Farming II," *Vermont Agricultural Experiment Station Bulletin 256* (Burlington: Vermont Agricultural Experiment Station, 1926), 19.
84 J. L. Hills, "Studies in Vermont Dairy Farming VI. The Position of Northern Vermont Among American Dairy Farming Regions," *Vermont Agricultural Experiment Station Bulletin 329* (Burlington: Vermont Agricultural Experiment Station, 1929).
85 Peet, "Problems of Land Utilization," 152.
86 One 1929 analysis of Marshfield and Cabot showed an average of forty-six acres in cropland, and another in 1927 showed an average of forty-three acres in Randolph and Royalton. Both these areas were successful dairy regions. H. P. Young, "Studies in Vermont Dairy Farming III," *Vermont Agricultural Experiment Station Bulletin 268* (Burlington: Vermont Agricultural Experiment Station, 1927), 10; E. W. Bell, "Studies in Vermont Dairy Farming V," *Vermont Agricultural Experiment Station Bulletin 304* (Burlington: Vermont Agricultural Experiment Station, 1929) 4.
87 H. P. Young, "Studies in Vermont Dairy Farming VII: Charlotte, Ferrisburg, and Panton Area," *Vermont Agricultural Experiment Station Bulletin 329* (Burlington: Vermont Agricultural Experiment Station 1931.)
88 Hooker, "Studies in Vermont Dairy Farming II," 27. The figures produced by other specialists in this series suggest that one farm operator working on his own could handle only "12 or 14 cows ... raising the crops therewith to feed them, keeping a small flock of hens, and an acre or two of cash crops on the side." Hitchcock, "A Study of Vermont Dairy Farming," 27.
89 Barron, *Those Who Stayed Behind*, 63.
90 Muse, "Standard of Living," 18.
91 Bell, "Studies in Vermont Dairy Farming V," 16.
92 Hitchcock, "A Study of Vermont Dairy Farming," 11.
93 Young, "Studies in Vermont Dairy Farming III," 2. In fact, almost every time an author reported these numbers, he also had to attach a disclaimer, as Joseph Hills did in his report on national statistics of profit and loss in dairying: "In not a single area did returns exceed gross costs"–but such numbers "do not possess great significance in any absolute sense." Hills, "Studies in Vermont Dairy Farming VI," 23.
94 S. W. Williams, "Studies in Vermont Dairy Farming VIII: Orleans, St. Albans, Randolph, and Richmond Areas," *Vermont Agricultural Experiment Station Bulletin 307* (Burlington: Vermont Agricultural Experiment Station 1934), 14, 34.

Chapter 6

1 Interview #50. Unless otherwise noted, all numbered interviews are from the Eugenics Survey of Vermont Migration Study field notes for Jamaica. VSARA; PRA-005; Container PRA-00014.
2 Interview #31.
3 Anderson came from a similarly tightknit rural community: what was at that time the small prairie town of Selkirk, Manitoba, part of a large settlement of Icelandic immigrants. She left there to attend the University of Manitoba in Winnipeg. Although she became a naturalized U.S. citizen and worked throughout her career in the U.S., she returned home late in life and was buried in Winnipeg.
4 Paul Searls, *Repeopling Vermont: The Paradox of Development in the Twentieth Century* (Barre, VT: Vermont Historical Society, 2023), 68.
5 Robert Alsop Riley, "Kinship Patterns in Londonderry, Vermont, 1772–1900," PhD dissertation, University of Massachusetts, Amherst, 1983, 169–170.
6 Interviews #94, #95, #96. "Old Wood place" is in *Brattleboro Reformer*, January 15, 1914.
7 The interview is listed under her name; it seems clear she owned the farm. Interview #88.
8 Interview #95.
9 The 1940 census asked where the person had been living in 1935 and whether they had been living on a farm.
10 Interview #94.
11 Interview #43.
12 Interview #22.
13 Interview #206A.
14 Interview #16.
15 In some ways the social world was quite similar to the one Hal Barron described in nineteenth-century Chelsea. Barron provides an evocative description, including both his appreciation of the richness of the social life and his own sense of profound relief that he himself was not required to participate in it. Barron, *Those Who Stayed Behind*, 123.
16 *Brattleboro Daily Reformer*, December 5, 1930.
17 Interview #29.
18 Interview #16.
19 Riley, "Kinship Patterns in Londonderry," 169–170.
20 Interview #206B.
21 Interview #129.
22 Warren E. Booker, *Historical Notes, Jamaica* (Brattleboro, VT: Hildreth Co., 1940), 101–104; 107–109.
23 Interview #44.
24 Interview #204.
25 Interviews #165; #231; #116; #23.
26 Interview #6.

27 Interview #31.
28 All the soloists lived in the central village, but perhaps some of the twenty members of the chorus were from outlying areas. Villagers typically had many more opportunities to participate in town events than did those in the outlying areas, if only for practical reasons. But Hal Barron notes, based on several sociological studies of upstate New York towns in the 1930s, that towns around the size of Chelsea and Jamaica (around five hundred people) encouraged the greatest inclusion of people outside the central village. Barron, *Those Who Stayed*, 129, note 41.
29 "Jamaica Christmas Church Services," *Brattleboro Daily Reformer*, December 19, 1930, 10. There seem to have been dozens of such "Christmas Cantatas" put together by different compilers. They were performed in many different towns in 1930.
30 "Jamaica Christmas Church Services," 10. Interviews #204; #129; #205; #221; #57.
31 Interviews #193A; #220.
32 *Brattleboro Daily Reformer*, May 19, 1928, 7.
33 "Old Home Day in Jamaica Today," *Brattleboro Daily Reformer*, August 7, 1930, 1.
34 Interview #204.
35 Interviews #223; #197; #196.
36 Interview #19.
37 Booker, *Historical Notes*, 128.
38 "Eureka Awakes" was the original title of a three-act play later retitled "Take My Advice." Eugene Hafer, *Take My Advice* (Boston: Walter H. Baker Co, 1924).
39 Booker, *Historical Notes*, 125.
40 Interview #56.
41 In the fall of 1930, for example, the *Reformer* reported that the West River Dramatic Club presented "Pat Piper's Place." The "Victorian Star chapter" served supper and the "Jolly Five of Jamaica furnished music for dancing." Three days later the P.T.A. in Rawsonville performed a play called "No Trespassing."
42 "Selective Migration from Three Rural Vermont Towns," *Fifth Annual Report of the Eugenics Survey of Vermont* (Burlington: Eugenics Survey of Vermont, 1931), 73.
43 The following material about town offices and government comes from the yearly town reports published by the town of Jamaica, available at Silver Special Collections Library at the University of Vermont. Six people served as listers over the fifteen years: one held the post for all the fifteen years between 1920 and 1935; another held the office for ten, and a third for nine. Similarly, fourteen people served as auditors: one for twelve years, one for six. The school board records are a little more confused because the town opted for a three-person board only in 1925, but a total of seven people served over those ten years: one woman held the office all ten years, and three others for five years each. In the town records collection held at the University of Vermont's Special Collections, the year 1926 is missing, so these figures assume continuity in that year.
44 Interviews #211, #96, #6.
45 Seven men served as listers between 1920 and 1935. Five were farmers, two were not, but the two who were not were in office only a total of four years. From 1925 on, all the listers were farmers.
46 The listers were also geographically distributed. The other boards operated differently: no geographical equity was attempted with the school directors; all but one of the auditors came from the village.

47 Interviews #18, #44.
48 Interview #44.
49 Mr. and Mrs. Tower did not stay in Jamaica, but they did not go very far away. In 1940 they were living in Brookline, three towns away, renting a farm and listing their occupation as "sugaring." In 1950 they were in Townshend, next door to Jamaica, but no longer on a farm. Mr. Tower was once again working in a sawmill; Mrs. Tower was working as a cook in a hospital, and the two daughters still at home were working in an office and a furniture factory.
50 Their last year in office was 1931; between 1920 and 1931 there was just one year when one of the family was not in charge of the roads.
51 *Annual Report of the Board of Officers of the Town of Jamaica, Vermont* (Rutland, VT: Tuttle, 1927).
52 One said that "only the men in the village get the road jobs," and the other complained that the town gave work to "single men or men who didn't need the money to work on its roads," instead of reserving the work for "married men with families." Interviews #220, #55.
53 That number included mostly road work wages, but also payments to elected officers, cemetery caretakers, people who cleaned the schoolhouses and brought them wood, and other such work.
54 Working on the roads was not funded entirely by the town, so this number does not account for everyone. After the flood of 1927, there was a major one-time repair project for the railroad, as well as repairs to bridges. Those projects were state-funded and sometimes paid higher wages for more skilled work over longer seasons. In 1930, they may have occupied more of the time of workers than did the usual town maintenance.
55 It included, for example, the retired dentist, the deputy sheriff, and a prosperous farmer.
56 Jamaica Town Auditor's Report, bound with Town Report, 1930.
57 "Selective Migration," 74. Interview #105.
58 Interview #206A and #182. There were no explanations of that exception—no loss of job or illness.
59 The paper used the names of the men hosting the hunters. I have not done so here because of their connections with the eugenics survey. *Brattleboro Daily Reformer*, November 13, 1930.
60 Muriel Follett, *New England Year: A Journal of Vermont Farm Life* (Brattleboro, VT: Stephen Daye Press, 1939), 187, 194.
61 Follett, *New England Year*, 192.
62 Follett, *New England Year*, 187.
63 Interviews #35; #92; #58.
64 Nicole Hahn Rafter, *White Trash: The Eugenic Family Studies, 1877–1919* (Boston: Northeastern University Press, 1988) 17.
65 Lewis Cecil Gray, *Introduction to Agricultural Economics* (New York: Macmillan, 1924), 14.
66 Interviews #97, #45.
67 Interview #117.
68 Interview #45.
69 Helen and Scott Nearing, *Living the Good Life: How to Live Sanely and Simply in a Troubled World* (New York: Schocken, 1970), 53.

70 Robert Frost, "A Time to Talk," *A Mountain Interval* (New York: Henry Holt, 1916), 44.
71 Helen and Scott Nearing, *Living the Good Life*, 53.
72 Interview #46.

Chapter 7

1 Parts of this chapter have appeared previously as "Vermont Becomes a Way of Life," *Vermont History* vol. 85, no. 1 (Winter/Spring 2017.) I thank the *Vermont History Journal* for permission to reprint sections of that work.
2 "Meeting of the Vermont Commission on Country Life, Van Ness House, October 9, 1929," 60. Minutes published in the VCCL's *News Letter* (December 1929), Silver Special Collections Library, University of Vermont.
3 Herbert Agar, "Introduction," *Free America* vol. 1, no. 1 (January 1937): 1.
4 Walter Crockett, "Will Vermont Look Forward?" *Driftwind* vol. 2, no. 1 (June 1927).
5 Arthur Patten Wallace, "'Progress' in Vermont," *Driftwind* vol. 2, no. 6 (May 1928): 6.
6 Vrest Orton, "Vermont—for Vermonters," *Driftwind* vol. 3, no. 1 (July 1928): 26.
7 Paul Prentiss Jones, "To your Tasks, Vermonters," *Driftwind* vol. 3, no.2 (September 1928). Jones had earned a B.A. from Dartmouth and an M.S. in agriculture at UVM before returning home to take over the family farm. *Brattleboro Daily Reformer*, May 24, 1945.
8 Coates continued to publish the anti-tourism tirades of writer and publisher Vrest Orton. Orton was a vocal critic of the progressive mindset of many of the leaders of the Commission: "Vermont's proper logical future does *not* lie in its development as a Summer Playground for the Nation," (a mocking reference to another state-generated slogan). Instead, he argued, "Vermont's proper and logical development" lay with its most traditional occupation: "in agriculture!" The next year, he published a sequel with an even sharper broadside. In "How to Make Vermont Free," Orton argued that the best way to protect Vermont from the ravages of modernity was for the state to secede from the Union. Vrest Orton, "How to Make Vermont Free (By a member of the Vigilance Committee)," *Driftwind* vol. 3, no. 4 (April 1929): 46. A final essay was titled: "Declaration of Independence for Vermont," *Driftwind* vol. 3, no. 5 (March 1929).
9 The degree to which Fisher also supported the Commission's eugenics agenda has been the subject of debate in Vermont recently, culminating in a successful effort to remove her name from the highly regarded Dorothy Canfield Fisher Book Award for children's literature. "Dorothy Canfield Fisher Book Award to Be Re-Named," *Seven Days*, Friday, May 3, 2019.
10 *Understood Betsy* reflects Fisher's embrace of Montessori educational theories, as several scholars have pointed out. See Elizabeth J. Wright, "Home Economics: Children, Consumption, and Montessori Education in Dorothy Canfield Fisher's Understood Betsy," *Children's Literature Association Quarterly* vol. 32, no. 3 (2007): 217–30.
11 For an introduction to Fisher's role as interpreter and promoter of the state, see Ida Washington, "Dorothy Canfield Fisher's *Tourists Accommodated* and Her Other Promotions of Vermont," *Vermont History* vol. 65, nos. 3 and 4 (Summer/Fall 1997).
12 Dorothy Canfield Fisher, "Vermont, Our Rich Little Poor State," *The Nation*, May 31, 1922, 643.

13 Percy Hutchinson, "Walter Hard's Pungent Poems of New Englanders," *New York Times*, September 10, 1939.
14 John Farrar was a native of Vermont but lived in New York at the time, where he had just co-founded the publishing company Farrar and Rinehart. Margaret Hard, *A Memory of Vermont: Our Life in the Johnny Appleseed Bookshop* (New York: Harcourt Brace, 1967), 141.
15 Hard's essay was published first in the *Survey Graphic*, a magazine designed to make the results of new sociological research accessible to a broad audience of non-specialists. In a few years, it was re-published for a much broader audience in *Reader's Digest*.
16 Walter Hard, "Vermont, A Way of Life," *Survey Graphic* 21, no. 4 (July 1932): 291.
17 Walter Hard, "Vermont ... to Me," *Vermont Life* vol. 1, no. 1 (Fall 1946), 14.
18 Dorothy Canfield Fisher, "Vermonters," *Vermont: A Guide to the Green Mountain State* (Boston: Houghton Mifflin, 1937), 7.
19 Hard, "Way of Life," 301–3.
20 Crane Brinton, review of Charles Edward Crane, *Let Me Show You Vermont* and Vrest Orton, *And So Goes Vermont*, in *New England Quarterly* vol. 11, no. 1: 197.
21 Elliott Merrick, *Green Mountain Farm* (New York: MacMillan Company, 1948), 9.
22 Ellen Lew Buell, "The New Pioneers," *New York Times*, September 16, 1934, 6.
23 The twelve-month period ending April 1, 1930, saw 680,422 Americans leave cities for farms—a net gain of 307,939 people on farms, and a 1 percent increase in the entire farm population in the five months following the stock market crash. *U.S. Census of Agriculture: 1935*, Volume 3: (Washington, DC: U.S. Gov. Printing Office, 1935),143.
24 The state showed an increase from 24,898 farms in 1930 to 27,061 in 1935. Windham County reported an increase from 1,665 to 1,914. In those same five years, 12,275 new people appeared on farms in the state, a 10 percent increase. Windham County saw a 12 percent increase in the farm population. These numbers were not unusually high in the context of New England: southern New England states gained many more back-to-the-land migrants in the early part of the Depression. The numbers quickly turned around in the period between 1935 and 1940, returning to the pattern of declining numbers of farms. *U.S. Agricultural Census: 1935*, Vol. 3, Table 6, 149.
25 Robert Frost, "Build Soil," *A Further Range* (New York: Henry Holt, 1936), 85–97.
26 Merrick, *From This Hill*, 29–30.
27 Dorothy Thompson, *Concerning Vermont* (Brattleboro, VT: Hildreth, 1937), 11–12.
28 Bernard DeVoto, "New England: There She Stands," *Harper's Magazine* (March 1932): 405–7.
29 DeVoto, "New England: There She Stands," 415.
30 The digital media in particular are filled with writers attempting to claim Frost's legacy for modern conservatism. See, for example: https://theimaginativeconservative.org/2022/01/robert-frost-imaginative-conservative-rv-young.html, https://chroniclesmagazine.org/web/robert-frost-social-and-political-conservative/.
31 Louis Masur, "Bernard DeVoto and the Making of the Year of Decision: 1846," *Reviews in American History* vol. 18, no. 3 (September 1990): 438.
32 DeVoto and Thompson were both sympathetic to a minor political movement called decentralism, which I have written about in *Back to the Land: The Enduring Dream*

of Self-Sufficiency in Modern America (Madison: Wisconsin University Press, 2011), and in "Vermont as a Way of Life," *Vermont History* vol. 85, no. 1 (Winter/Spring 2017): 43–64.

33 Roosevelt himself was deeply interested in the project of removing farmers from "submarginal" land as a means of stabilizing the agricultural economy. The New Deal plans emerged in large part from experimentation in New York in the years when he was governor of the state. William A. Hoglund, "Abandoned Farms and the 'New Agriculture' In New York State at the Beginning of the Twentieth Century," *New York History* vol. 34, no. 2(April 1953): 185–203.

34 Richard M. Judd, *The New Deal in Vermont: Its Impact and Aftermath* (New York: Garland, 1979), 91.

35 The story of the stand-off between federal agents and the forces gathered by Aiken is recounted in Sara Gregg, *Managing the Mountains: Land Use Planning, the New Deal, and the Creation of a Federal Landscape in Appalachia* (New Haven, CT: Yale University Press, 2010), chapter 6," Reforming Submarginal Lands."

36 George Aiken, "Not So Submarginal," in *Speaking from Vermont* (New York: Stokes and Co., 1938), 7–8.

37 Aiken, "Not So Submarginal," 7.

38 Aiken, "Not So Submarginal," 18–19.

39 In an exchange of letters with the director of the Vermont State Office of Publicity in January 1936, Aiken compared his count to theirs. The director reported he had counted a total of 110 sales of farms to out-of-staters in Windham County in 1935. George D. Aiken papers, Crate 60, Box 3, General Correspondence, 1936, Jan-Feb. 60-3-4. Silver Special Collections, University of Vermont Library.

40 Vermont Development Commission, *Graphic Survey, A First Step in State Planning for Vermont*, submitted to the Vermont State Planning Board and National Resources Board (Montpelier, 1935), XII.

41 As I noticed recently, Paul Searls found this same funny passage and uses it to the same effect in his study of Landgrove, a town just north of Jamaica. Paul Searls, *Repeopling Vermont: The Paradox of Development in the Twentieth Century* (Barre: Vermont Historical Society Press, 2019), 140.

42 The figure seems to include both work relief programs and direct relief of cash, food, or shelter.

43 DeVoto's summer residence of Morgan reported somewhere between 0.1 and 2.4 percent on relief, and several tiny surrounding towns reported none. A third conspicuous area was the mountain towns west of Jamaica in Bennington County.

44 Aiken, "The Tragedy of Relief," in *Speaking from Vermont*, 195.

45 The hardest-hit farmers in Maine, for example, were in Aroostook County, where many farmers were wholly dependent on the commercial potato crop. Richard Condon, "Nearing the End," *Maine History* vol. 31, no. 3 (1992): 142–73. Among the northern New England states, Vermont was by far the most dependent on the marketing of milk. In 1930, only 17 percent of Maine's farms and 29 percent of New Hampshire's were dairy farms, while in Vermont, 58 percent of farms specialized in milk.

46 In 1920, farmers had received a wholesale average of $3.22 per one hundred pounds of whole milk. By 1932, that number had dropped to $1.28. In 1939, the price was still only back up to $1.69. Shepard B. Clough and Lorna Quimby, "Peacham, Vermont: Fifty Years of Economic and Social Change," *Vermont History* vol. 51, no. 1 (Winter, 1983): 11–12.

47 S. W. Williams, "Studies in Vermont Dairy Farming: VIII. Orleans, St. Albans, Randolph and Richmond Areas," *Vermont Agricultural Experiment Station Bulletin* 307 (Burlington: Vermont Agricultural Experiment Station, 1934), 25.

48 John A. Hitchcock and S. W. Williams, "Studies in Vermont Dairy Farming: IX. The Champlain Valley During a Major Depression," *Vermont Agricultural Experiment Station Bulletin* 405 (Burlington: Vermont Agricultural Experiment Station, 1936), 23.

49 In 1935, only 34 percent sold enough dairy products to be categorized as dairy farms. H. R. Varney, *Economic Facts: Fourteen Counties* (Burlington: Vermont Extension Service, 1935).

50 In Maine, farmers experienced similar problems, as farmers typically worked in canneries, fishing, or logging part-time, and those jobs all failed. Condon, "Nearing the End," 147–8.

51 Harry C. Woodworth, Max F. Abell, John C. Holmes, *Land Utilization in New Hampshire: I. Problems in the Back Highland Areas of Southern Grafton County*, (Durham: University of New Hampshire, 1937), 54. Woodworth wrote a somewhat less sympathetic account in the *Journal of Farm Economics*, but there too he pointed to the fact that "certain individuals" were unable to adjust to a commercial life and "should not be denied the opportunities of a self-sufficing mode of living," with the important proviso that they "choose locations where the social costs in roads and schools are not prohibitive." H. C. Woodworth, "A Century of Adjustments in a New Hampshire Back Area," *Journal of Farm Economics* vol. 19, no. 2 (May 1937): 237.

52 Woodworth, "Land Utilization in New Hampshire, "58.

53 New York State College of Agriculture and Life Sciences, T. E. LaMont, *The People in Land Classes I and II* (Ithaca, NY, 1940). This is a summary of results from studies conducted between 1927 and 1938.

54 This was a study completed before the Depression, but Muse probably wrote the report at a point when the impact of the Depression was becoming evident. Muse, Marianne, "Standard of Living on Specific Owner-Operated Vermont Farms," *Vermont Agricultural Experiment Station Bulletin* 340 (Burlington: Vermont Agricultural Experiment Station, 1932), 24.

55 Dorothy Emery, "A Food Budget for Vermont Farm Families," *Vermont Agricultural Experiment Station Bulletin* 393 (Burlington: Vermont Agricultural Experiment Station, 1935), 4.

56 Emery, "Food Budget," 4–8.

57 She listed apples, blackberries and raspberries (both wild and cultivated), wild blueberries, cherries, currants, gooseberries, grapes, pears, plums, and strawberries.

58 There were equally long lists of things farm families should buy, ranging from chocolate and coffee to oats and peanut butter.

59 Ogden later recalled remodeling houses or barns as homes for the documentary filmmaker Robert Flaherty, the artist Bernadine Custer, and the violinist Nathan Milstein, among many others. Samuel Ogden, *The Cheese that Changed Many Lives:*

Or, A Sentimental History of a Tiny Town High in the Green Mountains (Landgrove, VT: Just-So Press, 1978), 92–98. Paul Searls recounts Ogden's Landgrove ventures in *Repeopling Vermont* (2019)

60 The Stephen Daye Press also published collections of Vermont biography, poetry, and prose, and the first volume of what would become Helen Hartness Flanders's definitive collection of Vermont and New England folk music.
61 Muriel Follett, *New England Year* (Brattleboro, VT: Stephen Daye Press), 51.
62 Follett, *New England Year*, 112.
63 Follett, *New England Year*, 125.
64 Follett, *New England Year*, 79.
65 Perhaps Anderson herself would not have seen the contradiction here, since the Folletts were so clearly middle class, but this was precisely the kind of control they were reluctant to see in the hands of less prosperous households in Jamaica.
66 Follett, *New England Year*, 84.

Epilogue

1 A number of farm removal programs were carried out by the various New Deal agencies in other parts of the country. One example of large-scale removal took place in the Blue Ridge Mountains, as Sara Gregg has described in *Managing the Mountains: Land Use Planning, the New Deal, and the Creation of a Federal Landscape in Appalachia* (2013).
2 University of Vermont Extension Service, *Farm Census for the Towns in Vermont Based on the Bureau of Census Unpublished Data, January 1, 1945* (Burlington, VT: Extension Service, 1946). Typescript unpaginated.
3 By the count of the 1950 census, there were forty-one farming-related jobs, including hired men, cattle dealers, sugar makers, and farmers. Fifty men reported jobs in construction, in woods and logging, and in work on the roads.
4 *Farm Census*, 1946.
5 Out of 158 "heads of household," 51 arrived between 1930 and 1940. There were also thirty-five men and women who did not count as "heads of household" in this census because they worked at or boarded in someone else's household. That included a variety of mostly single adults: housekeepers, mill workers, hired men, and loggers. They were a more highly mobile population than the rest: among the thirty-five, only six had been in Jamaica in 1930—83 percent had moved to town in the past decade. Counting those extra adult individuals, a total of 41 percent of Jamaica's households had come to town since 1930 (80/193).
6 For this number, I included the thirty-five men and women who were not officially "heads of household," to capture a clearer picture of migration in those few years. A total of 51 out of 193 had moved to Jamaica in the previous five years: 26 percent.
7 Eighteen (35 percent) from nearby small towns; twelve from the rest of Vermont, New Hampshire, and Maine (24 percent); twenty-one (41 percent) from Massachusetts, Connecticut, New York, Pennsylvania, New Jersey.
8 "Bauer" is the last pseudonym I will use: all family names after this point are the real ones given in the records. I have provided pseudonyms only for people whose families were interviewed by the Eugenics Survey.

9 As young adults back in 1900 they had lived together in New York City, where he had worked as a reporter and she as a bookkeeper. Mabel had returned home to Illinois to care for their aging parents, and Frederick had become a successful author of detective fiction: https://jiescribano.wordpress.com/2020/05/22/frederick-irving-anderson/. Blog: "A Crime is Afoot," Jose Ignacio Escribano.
10 As the census records show, Seaman worked his way up from a listing as a carpenter in 1920 to a building inspector in 1930, and then to his work as a real estate appraiser.
11 Frederick Anderson listed no occupation, but back in the 1910s he had written two books about farming and technology, so perhaps he still retained some interest in the subject: *The Farmer of Tomorrow* (1913) and *Electricity for the Farm* (1915).
12 Both the woods and the farm were an important "draw" for teachers and students. For example, David Benton Crittenden, a Yale-trained architect, moved north to teach students mechanics and blacksmithing. Decades later, his obituary reported that he had "enjoyed forestry, working on the land and growing apples, raspberries and blueberries." *Hartford Courant*, April 6, 2001, 284. Information on the farm's history comes from the obituary of David Newton in the *Brattleboro Reformer*, February 25, 1982; and the obituary of his son John Newton in the *Rutland Herald*, February 13, 2013. The *Princeton Alumni Weekly* also ran a paragraph about the school. *Princeton Alumni Weekly* 46, December 14, 1945, 17. For an account of the long-term impact of the children of the school's founders on local education, see *Windham News and Notes* Vol. VI, no. 5 (March/April 2009): 4.
13 For Helen Knothe Nearing, see Margaret Killinger, *The Good Life of Helen K. Nearing* (Hanover, NH: University Press of New England, 2007.) For Scott Nearing, see John A. Saltmarsh, *Scott Nearing: The Making of a Homesteader* (Philadelphia: Temple University Press, 1991); and Stephen J. Whitfield, *Scott Nearing: Apostle of American Radicalism* (New York: Columbia University Press, 1974). For Scott's own version, see Scott Nearing, *The Making of a Radical: A Political Autobiography* (New York: Harper and Row, 1972), and *Man's Search for the Good Life* (1954; repr., Harborside, ME: Social Science Institute, 1974). And for Helen's own version, *Loving and Leaving the Good Life* (White River Junction, VT: Chelsea Green Press, 1992).
14 The Nearings tell that part of their story in *The Maple Sugar Book* (White River Junction, VT: Chelsea Green Press, 2000). In this updated edition, historian Greg Joly adds valuable context, tracing the Nearings' relationships with their co-workers and their community in Jamaica and Winhall.
15 Distributism is a political and ideological movement that has its roots in early twentieth-century England, difficult to pin down, but often combining decentralist and sometimes anarchist beliefs in the widespread "distribution" of property with more socially conservative views about the family, community, and religion. It was associated in this generation of Americans with the work of Ralph Borsodi.
16 See Greg Joly and Rebecca Lepkoff, *Almost Utopia: The Residents and Radicals of Pikes Falls, Vermont, 1950* (Montpelier: Vermont Historical Society, 2008), 41–66. Lepkoff's photographs date from the 1950s. Combined with Joly's extensive research and clear writing, they present a vivid portrait of these communities and their relations with their neighbors. The stories of two draft resisters, Lowell Naeve and Norman Williams, appear on pages 53–57.
17 "They Get Away from it All in Pikes Falls," *Brattleboro Daily Reformer*, November 10, 1948, 9.

18 They certainly had their opportunities: the article featured a photograph of handsome young Norman Williams in a flannel shirt, next to another photo of the stone house he had built for his family. The article did not mention that Williams and his wife abhorred the "good war" the United States had just won. Nor did the writer refer to his 4-F draft status on account of his being classified as "insane" because of his opposition to that war. Joly and Lepkoff, *Almost Utopia*, 57.
19 "Fugitives from Reality," *Brattleboro Daily Reformer*, November 17, 1948, 4.
20 Joly and Lepkoff, *Almost Utopia*, 58.
21 The Vermont Historical Society has collected a wide-ranging series of oral histories with residents who moved to Vermont during the 1960s and 1970s, illustrating the long-lasting and far-reaching impact of that migration on the state. https://www.digitalvermont.org/vt70s.
22 The leading figure in the development of organic gardening and farming in the United States was J. I. Rodale, who began publishing the magazine *Organic Farming and Gardening* in 1942.
23 Organic practices were often associated with Quakers and other pacifists; followers of eastern religions like Helen Knothe Nearing; and with Ralph Borsodi's distributist movement, as expressed in the magazine called at various times the *Interpreter*, *One Way Out*, and *Green Revolution*. For a description of those movements in Jamaica, see Lepkoff and Joly, *Almost Utopia*. For an account of the political ideologies of the back-to-the-landers of the twentieth century, see Dona Brown, *Back to the Land: The Enduring Dream of Self-Sufficiency in Modern America* (2011).
24 See, for example, "Oral history interview with Don Hooper," *Digital Vermont: A Project of the Vermont Historical Society*, accessed August 10, 2024, https://www.digitalvermont.org/vt70s/AudioFile1970s-12.
25 Sabra Field moved to Vermont in 1967, Mary Azarian in 1963; Woody Jackson began painting in Vermont in 1972.
26 https://savingplaces.org/11most-past-listings.
27 In 2022, spiraling costs for feed and fuel combined with inelastic prices for milk to make the situation desperate for the rapidly dwindling number of dairy farms in the state. That was followed by massive flooding in 2023 and 2024.
28 https://www.vermontpublic.org/show/made-here/2024-03-14/10-years-after-hide-migrant-farmworkers-in-vt-are-still-seeking-rights. Derek Brouwer, "Migrant Workers Hold Up Vermont's Dairy Industry—and Are Fighting for Better Working Conditions," *Seven Days*, May 31, 2023.
29 https://www.vermontpublic.org/local-news/2022-08-16/food-insecurity-in-vermont-reached-all-time-highs-during-the-pandemic-then-came-inflation.
30 "Small Axe Farm's Evolution from Homestead to No-Till Farm," https://www.mofga.org/stories/community/small-axe-farm-no-till/. The name of the farm refers to pioneering reggae singer Bob Marley's 1973 song, "Small Axe," an anthem celebrating the power of the "small axe" against the "big tree."
31 https://smallaxefarm.com.
32 "Kearsarge Gorge Farm," Northeast Organic Farmers Association, New Hampshire, https://www.youtube.com/watch?v=GzpJY34wHso.

Index

accounting systems: of agricultural economists, 149–150; of Cornwall, Jamaica, and Waitsfield farmers, 110–13
agrarianism, 42, 83, 224n2, 224n3, 225n18
agricultural commissioner, 48, 50–52, 226n22. *See also* Jones, Edward H.
agricultural economics: dairy studies at Vermont Agricultural Experiment Station, 50, 147–150; field of, 47, 51–52, 171; influence on Vermont Commission on Country Life, 57–58
Agricultural Experiment Station, Vermont, 49–50, 54, 61. *See also* agricultural economics; home economics
Agricultural Extension Service, Vermont, 50, 74, 187
Aiken, George, 12, 186–90, 192, 197
ambition, lack of (as eugenicist critique), 74–79, 81–84, 86, 116, 231n34. *See also* contentment, laziness
Anderson, Elin (Migration Study lead investigator), 69, 75, 80–81, 232n36, 244n3; analysis of Jamaica, 9, 73–81; *We Americans*, 9, 73, 230n8. *See also* Migration Study
automobiles, 104, 133–34

back-to-the-land migration: to Jamaica, 202–204; and radical politics, 205–7; to Vermont, 181–83, 248n24
Bailey, Liberty Hyde, 46–47
Barnard, Vermont, 185, 228n59. *See also* Thirteen-Town Study
Barron, Hal (*Those Who Stayed Behind*), 30, 64, 148, 213n2, 224n6, 245n28. *See also* Chelsea, Vermont

Board of Agriculture, Vermont, 4, 16, 43–46, 53–56, 83, 226n22; membership of, 49–50; replacement by commissioner, 48–52
Bradlee, Thomas: director of Vermont Extension Service, 58; member of Vermont Commission on Country Life, 74, 230n21
Brighton, Massachusetts, meat market, 20, 33
buckwheat, 37, 223n91
"Build Soil," (Frost), 13
butter, *see* cream

capital (required for farming), 22, 82, 104–105, 116–17
cattle dealers, Jamaica, 32–36, 110, 201–2
cattle trade, 20–21, 32–36, 202; 218n35. *See also* Brighton, Massachusetts, meat market; cattle dealers
cattle: in Cornwall and Waitsfield, 108–9; in Jamaica, 20, 26–28, 37, 108–10; in New England, 20, 218n28
census: advice to enumerators, 112–13, 126; count of farms in Jamaica, 125–26, 238n79; farm categories used in 1930, 107, 235n24; nineteenth-century agricultural, 221n68
Chelsea, Vermont: community compared to Jamaica, 244n15, 245n28; farms compared to Jamaica farms, 29, 30, 221n61, 224n6, 234n4, 241n34; compared to towns of Thirteen-Town Study, 64. *See also* Barron, Hal
Choate, Marjorie (Migration Study investigator), 1–2, 69, 80–81, 229n3. *See also* Migration Study

255

Clayton, Claude F. (agricultural economist), 61, 64, 132
Coates, Walter (editor, *Driftwind*), 176–77, 247n8
community (as reason for remaining or returning to Jamaica), 86, 88, 154–58, 160–64
community organizations, Jamaica: Benefit Club, 159, 162–63; Masons, 159; P.T.A., 158; Women's Christian Temperance Union, 158–59
Concerning Vermont (Thompson), 185
contentment: as eugenicist critique, 76, 81–85, 95, 180; as virtue, 83, 97, 178–81, 197–200. *See also* ambition, laziness
conveniences, modern, 39, 133–35, 240n26, 240n27. *See also* automobiles, electricity, plumbing, radios, telephones
cordwood (as farm crop), 31, 36, 222n74. *See also* logging
corn, 31, 62, 103, 105, 221n70
Cornell University, New York State College of Agriculture, 46–48, 147, 225n21
Cornwall, Vermont: classification as "fair" or "high average" town, 5, 64, 74, 76; farms of, 107–9, 113, 119–23, 127–28; incomes in, 127–28; interpretation by investigators, 76, 85, 235n12, 239n3, 235n12; opportunities in, 135–40, 241n47; reasons for population persistence in, 86–90, 94, 233n60, 233n63
cost of living, Jamaica, 138–39. *See also* standard of living
Country Life Commission (national), 46–47
country life movement, 46–48, 225n18, 228n69
cows, milk: numbers in Jamaica, 30, 32; numbers in Vermont, 32. *See also* cattle, cream, milk
cream: cheese and butter production, 28–30; competition from western producers, 63; shift to fluid milk production from, 62–63, 99, 105, 107, 109, 123–24. *See also* milk
creameries, 63, 123–24

Crockett, Walter Hill (member of Vermont Commission on Country Life), 74, 176, 230n21

decline, rural: attributed to economic competition, 46, 58; attributed to inefficient farm practices, 47, 52, 58; attributed to loss of "best and brightest," 4, 72, 73, 75, 176; attributed to racial deterioration, 71–73; as diagnosed by Migration Study, 76–77, 81–84, 180; historiography of, 224n6; in New England, 2, 41–42, 83
deer hunting, 168–71, 173
deforestation: cattle and, 26–28; logging and, 28, 39, 64; measurement by census "improved" land figures, 27–28, 220n59; sheep and, 25–28. *See also* reforestation
Depression, Great: impact on dairy farms, 150, 192–93; influence on agricultural experts' advice, 192–97, 201, 250n51; influence on ideas about hill farms, 11, 174–75, 179–86, 189; migration to Jamaica during, 12, 202–5; migration to Vermont during, 139, 181, 183, 189, 248n24. *See also* New Deal
DeVoto, Bernard, 185–86
Diets, farm: food produced for home use, 119–22, 132–33, 182–83, 185; home economics studies of, 119–20, 129–33, 195–96, 240n21
Donahue, Brian (*The Great Meadow*), 215n4, 215n2, 218n28
Driftwind (Coates, editor), 176–77, 247n8
drovers, *see* cattle dealers; cattle trade

education, 136–38
electricity, 39, 133–35, 240n26, 240n27
Emery, Dorothy (home economist), 196, 223n89
emigration: from Cornwall and Waitsfield, 141–42; for education, 136–38; from Jamaica to adjoining towns, 91–92; from Jamaica to more distant locations, 92–93; from Jamaica and returning, 91–95, 142; to mills and workshops, 141–45

Enosburg, Vermont (subject of dairy study), 148
Eugenics Survey of Vermont, 2–6, 66–68, 69–72, 214n11; attempts to improve image of, 71–73, 76; studies of "defect, degeneracy, and dependence," 70–71; focus on white "Yankee" population, 3–4, 61, 72–74, 77, 213n5, 214n6; relationship to Vermont Commission on Country Life, 61, 66, 70–71, 229n10. *See also* Migration Study; Perkins, Henry

family: as reason for remaining in or returning to Jamaica, 86–89, 139–41, 153–156; as source of economic assistance, 87–89, 139–41. *See also* kinship networks
Farm Production and Marketing Committee, Vermont Commission on Country Life, 56–57
farms, dairy: Agricultural Experiment Station studies of, 50, 147–50; as census category, 107–8; in Chelsea, Vermont, 29; in Cornwall and Waitsfield, 108–10; decline of in twenty-first century, 209–10; expert preference for, 62–63, 147–50; impact of Depression on, 192–93; in Jamaica, 29–30, 34, 107–10; and labor needs, 30, 148, 150; pressure to modernize, 30, 33, 105; transition from wool, 29–30, 32
farms, general (census category), 107–8
farms, mixed: combining livestock, 27, 29; combining market and subsistence crops, 18–19, 29, 35–37, 99–100, 106–8, 110; modern diversified, 146–47, 198–99. *See also* farming, general (census category)
farms, numbers of: in Jamaica, 39, 100–101, 201–2; in hill towns, 62, 146–47; in valley commercial farming regions, 209–10
farms, self-sufficing (census category), 107–9, 118, 120, 123, 193. *See also* farms, subsistence; home use (eugenics category)

farms, subsistence, 9, 12, 120–23, 184, 193; combined with off-farm work, 118–19, 145, 150–51; expert opinions of, 118–24, 194–96, 250n51; as family strategy, 97, 118–19, 155–56. *See also* home use (eugenicist category); farms, self-sufficing (census category)
Fayston, Vermont, 228n59. *See also* Thirteen-Town Study
ferning (picking ferns), 110, 115, 137, 171
fertilizer, 105, 222n72, 235n19
Fisher, Dorothy Canfield, 12, 177–81; *Understood Betsy*, 178, 180; and Vermont Commission on Country Life, 178, 247n9
Fitzgerald, Deborah (*Every Farm a Factory*), 8
flood of 1927, 115–16, 246n54
Follett, Muriel, 169–70, 197–200
Forestry Committee, Vermont Commission on Country Life, 60, 65, 77–78, 187
forestry, discipline of, 48–50, 226n22
Frazier, William (Religious Forces Committee, Vermont Commission on Country Life), 74–75, 230n21
freedom: as feature of hill farm life, 60, 168, 175, 188, 199, 207; as reason to remain in or return to Jamaica, 95, 168–74, 242n68. *See also* Aiken, George; Nearing, Scott and Helen
Fritz, Jay J., (Forestry Committee, Vermont Commission on Country Life), 60
From This Hill Look Down (Merrick), 181–82, 184–85, 197
Frost, Robert, 13, 183–84; "Build Soil," 184; "A Time to Talk," 172–73

gardens, 120–22, 199; organic, 208–9, 253n23
Goshen, Vermont, 228n59. *See also* Thirteen-Town Study
Granville, Vermont, 228n59. *See also* Thirteen-Town Study
Graphic Survey: A First Step in State Planning for Vermont (Vermont State Planning Board), 190–92

Gray, L. C. (agricultural economist), 51–52, 171, 226n32
graziers, *see* cattle dealers, cattle trade
grazing: mixed cattle and sheep, 27, 220n58; among sugar maples, 101; suitability of northern hills for, 20–21, 54, 225n9; in woodlands, 33, 222n81, 234n10
Gregg, Richard (*The Power of Nonviolence*), 206–8
Gregg, Sara M. (*Managing the Mountains*), 225n19; 226n25, 227n45, 230n15, 231n22, 251n1

Hard, Walter, 179–81
Harris, Christopher ("The Road Less Traveled By"), 214n9, 236n46
Harvard Forest dioramas, 18, 25–26, 217n17
health (as reason for migrating to, remaining in, or returning to Jamaica), 88–90, 139, 143
Hill Country of Northern New England, The (Wilson), 215n1, 218n39, 224n6
hill farms: defenders of, 60–61, 175, 186–90; meaning of term, 4–5, 51–52, 210–11; nineteenth-century views of, 4–5, 43–46; twenty-first century types of, 210–12
hill towns, political representation of, 49, 66, 73, 226n25
Holstein cattle: as part of modernizing dairy production, 107; as symbol of Vermont, 29, 209, 222n85
home economics (Agricultural Experiment Station): standard of living studies, 129–36, 239n10; studies of farm access to modern technologies, 133–34; studies of farm diets, 119–20, 132–33, 195–96; studies of food self-sufficiency, 195–96; studies of importance of non-cash farm income, 130–33; studies of women's earnings, 128. *See also* Muse, Marianne; standard of living
home use, farming for (eugenicist category), 96–97, 118–24, 155, 238n72. *See also* farms, self-sufficing (census category); farms, subsistence

Human Factor, Committee on the, Vermont Commission on Country Life, 2, 61, 66, 70, 71, 78, 231n31

incomes: in Cornwall and Waitsfield, 127–28; in hill towns, 127, 132, 238n1; in Jamaica, 103–4, 127–28; non-cash, 123, 130–33; women's, 128–29. *See also* accounting systems; poverty; standard of living; wages

Jones, Edward. H. (Agricultural Commissioner), 51–52, 56, 74, 239n7
Judd, Richard (*Second Nature*), 216n3, 242n68

kinship networks: in Landgrove, Vermont, 154; in Londonderry, Vermont, 154; as reason for remaining in or returning to Jamaica, 154–55; relationship to "home use" farming practices, 155–56. *See also* family; status, social

labor shortages, rural, 35, 37; as barrier to dairy farming, 30, 147–48, 150, 222n79; as limit on maple production, 31–32
Land Utilization Committee, Vermont Commission on Country Life, 58–61, 65–66, 77–78
land utilization studies: New Hampshire, 194, 250n51; Vermont, *see* Thirteen-Town Study
Landgrove, Vermont, 154, 197, 215n16
laziness (as eugenicist critique), 81–83, 168–71, 199. *See also* ambition, contentment
literacy, 112, 236n40
logging: on farmers' own woodlots, 39, 64, 92, 123, 202; as off-farm work, 64, 114, 136, 144, 169, 242n67, 251n3; in Winhall, Vermont, 205. *See also* work, off-farm
Londonderry, Vermont, 140–41, 154
lumber industry: as intended beneficiary of farm removal proposals, 65–66, 77; in Jamaica, 7, 28, 39

manufacturing: in Chelsea, Vermont, 241n34; in Jamaica, 39, 41, 100, 136; in

lumber mills and sawmills, 64, 114–15, 122, 145, 234n3. *See also* logging
manure, 31, 221n71
maple sugar and syrup, 32–33, 123, 198, 202, 205, 222n75, 222n76, 222n80
McReynolds, Samuel ("Eugenics and Rural Development"), 228n69
Merrick, Elliot, 181–84, 197
Migration Study, 66–68; choices of three towns for study, 3–4, 73–75; documents generated by, 5–7, 214n11; focus on "Yankee" population, 71–73; non-Yankee population in Jamaica, 72, 230n13; reactions in Jamaica, 76; relationship to Eugenics Survey, 70–71; relationship to Vermont Commission on Country Life, 70, 73–75; research techniques, 5, 79–80, 85–86, 96–98, 102–3, 152–53, 157–58, 231n35; staff, 69–70, 80–81. *See also* Anderson, Elin; Choate, Marjorie; Perkins, Henry; Rome, Anna
migration to Jamaica, 93–94, 174, 248n24; reasons for, 85–90, 202–7
milk production: encouraged by agricultural experts, 147–49, 192–93; for household consumption, 20, 119, 196; infrastructure demands of, 63, 104–5; as modernizing step, 28–29, 105, 107–9; prices of, 148–49, 192–93; shift from cream to fluid milk, 62–63, 99, 105, 107, 109, 123–24, 235n27. *See also* cream
mill towns, migration from Jamaica to, 142–45
Moody, John (chair of Committee on the Human Factor), 60, 66, 231n31
Morgan, Vermont, 185, 192, 249n43
Mt. Holly, Vermont, 228n59. *See also* Thirteen-Town Study
Muse, Marianne (Agricultural Experiment Station home economist), 129–34, 148, 195–96, 237n66, 237n69, 240n21, 241n58. *See also* home economics; standard of living

Nearing, Scott and Helen, 12, 171–73, 205–7
New Deal, 11–12; distribution of "relief" programs in Vermont, 190–92; submarginal buyout plan, 12, 186–89, 227n45, 249n33, 226n25, 227n45, 230n15
New England Year (Follett), account of deer hunting, 169–70; account of farm life, 197–200
New England: agriculture in, 20, 25–26, 63, 218n28, 218n31; migrants to Jamaica, 93–94, 202–4; rural decline in, 2, 41–2, 83; southern settlement of northern, 14–16, 19, 216n3; trade between southern and northern, 20–21, 33–34, 218n35; "Yankee" ancestors of, 50, 72, 159. *See also* settlement
nutrition, *see* diet

off-farm labor, *see* work, off-farm
Ogden, Samuel, 197, 250n59
Orton, Vrest, 197, 247n8
Osterud, Grey ("Farm Crisis and Rural Revitalization in South-Central New York"), 215n16

pacifists, migration to Jamaica, 206–7
pageant, Jamaica, 161–4
Peach, Arthur W. (chair of Committee on Traditions and Ideals), 60–61, 175, 207, 228n56
Peet, Lemuel J. (agricultural economist), 61–65, 74, 111. *See also* Thirteen-Town Study
Perkins, Henry (director of Vermont Eugenics Survey), 2–3, 66, 70–72, 73, 214n6. *See also* Eugenics Survey of Vermont, Migration Study, Vermont Commission on Country Life
Petersham, Massachusetts, 16–19, 26, 48, 227n43
Phillips, Sarah T. (*This Land, This Nation*), 9–10
Pikes Falls, Vermont, 163, 170, 205–7, 252n16
Pittsfield, Vermont, 228n59. *See also* Thirteen-Town Study
Planning Board, Vermont State, 190–92
plumbing, indoor, 133–35, 153, 240n25, 240n27
Plymouth, Vermont, 228n59. *See also* Thirteen-Town Study

population loss, rural: in hill towns, 62, 138; in Jamaica, 39, 41, 100; in New England, 41–42; in New York, 46; in the United States, 41–42; in Vermont, 42–43; rebound in late twentieth century, 202–3, 208; reversal during Depression, 183, 189, 249n39
population persistence: of farm inheritors, 64, 146; reasons reported in Migration Study interviews, 85–95, 138–46, 158–74; of young Jamaicans, 84–85. *See also* community; family; freedom; health; kinship networks; retirement; security; status, social
potash, 18–19
potatoes: as crop, 31, 120, 188, 235n19, 221n69; as dietary staple, 132, 223n89, 240n21
poverty: in hill towns, 62, 64, 127, 188; in Jamaica, Cornwall, and Waitsfield, 127–28. *See also* incomes, standard of living
Power of Nonviolence, The (Gregg), 206
"Problems of Land Utilization in the Hill Towns of Vermont, Based on a Study of Thirteen Towns in 1929" (Peet). *See also*, Thirteen-Town Study
Proctor, Mortimer: as chair of Forestry Committee, 77; as governor, 226n22
profit and loss, *see* accounting systems
Putney, Vermont, 12, 20, 188, 191–92, 209

radios, 134, 135, 240n27
Rafter, Nicole Hahn (*White Trash*), 230n11; 232n38
railroad: and cattle trade, 33–35; and dairy products, 63; and lumber trade, 39; repair of, 115, 246n54
reforestation, 27–28, 100–102, 202, 234n6; as state policy, 60, 65, 101
removal (of population in hill towns), 5, 9, 59, 64, 66, 77–79, 186–88, 201
retirement (as reason for migrating to Jamaica), 88–89, 139, 241n47
Riley, Robert ("Kinship Patterns in Londonderry, Vermont"), 154, 158
Ripton, Vermont, 3, 172, 183–84, 228n59. *See also* Thirteen-Town Study
road work, 115–16, 144, 166–67; experts' disapproval of, 115–17

Rome, Anna (Migration Study investigator), 1–2, 69–70, 80–81, 84–85
Roxbury, Vermont, 228n59. *See also* Thirteen-Town Study
rural decline, *see* decline, rural
rural surveys, *see* surveys, rural
Rural Vermont: A Program for the Future (Vermont Commission on Country Life), 71, 78–79

Searls, Paul (*Repeopling Vermont*), 154, 215n16, 226,
security: as attributed to life on hill farms, 11, 42, 61, 175, 182–83; as reason for remaining in or returning to Jamaica, 144–46, 156
"Selective Migration from Three Rural Vermont Towns," *Fifth Annual Report of the Eugenics Survey of Vermont*, 77. *See also* condensed version in *Rural Vermont: A Program for the Future*, 78–79
settlement (of northern hills): attractiveness to southern New England farmers, 15–16, 19–20, 216n3; clearance process, 16–19; division of land, 16–17, 216n9, 216n10; historiography of, 14, 215n1, 216n2, 216n3
sheep: environmental impact of, 25–28, 220n54; grazing with cattle, 26–27, 220n58; numbers in Jamaica, 22–27; numbers in Vermont, 21–22, 23–24, 218n40, 219n47, 221n61. *See also* wool trade
Sherburne, Vermont, 228n59. *See also* Thirteen-Town Study
Shoreham, Vermont, 22, 24, 219n45
Shrewsbury, Vermont, 228n59. *See also* Thirteen-Town Study
skiing, 14, 91, 207–8
Small Axe Farm, 211
social status, *see* status, social
Soil Survey (Reconnaissance) of Vermont (Latimer), 54–55, 101
soil surveys, 52–55; conducted by national Bureau of Soils, 48, 53
soils, quality of: as judged by Aiken, George, 188; as judged by Board of Agriculture leaders, 43–45, 225n9,

217n24; as judged by early agricultural advisors, 15–16, 19–20, 29; as judged by residents of Jamaica, 82, 102–4; as judged by soil scientists, 52–55, 56, 103–5, 227n40, 227n41, 227n43; as unrelated to yield or profits, 62–63, 103–4. *See also* soil surveys, Thirteen-Town Study

Some Vermonters (Hard), 179

Speaking from Vermont (Aiken), 12, 188, 192

Spear, Victor (Board of Agriculture member), 48–49

standard of living: definitions of, 129–35, 239n10; education as factor, 136; in hill towns, 62, 127; housing as factor, 130; modern conveniences as factors, 133–35; non-cash income as factor, 130–33. *See also* home economists; incomes; Muse, Marianne; poverty

status, social: in Londonderry, Vermont, 158; Migration Study ranking system, 157–58; as reason for remaining in or returning to Jamaica, 158–68; related to kinship networks, 153; related to long family residence, 153, 158–60; related to membership in town social organizations, 158–64. *See also* community; family; kinship networks

Stephen Daye Press, 197

Stockbridge, Vermont, 228n59. *See also* Thirteen-Town Study

Stratton, Vermont, 20, 91, 170, 191, 207, 218n40

submarginal buyout plan, 12, 186–89, 227n45, 249n33, 226n25, 227n45, 230n15

submarginal (term coined by agricultural economists), 55–56, 77–78, 186

subsistence farms, *see* farms, subsistence

surveys, rural, 2, 61, 213n2; as developed at Cornell Agricultural College, 47, 50, 225n21

Taylor, Henry C. (director of Vermont Commission on Country Life), 47, 57

telephones, 39, 134, 240n26

Thirteen-Town Study, 61–65; on emigration from hill towns, 138, 142, 146; on farming practices, 103–4, 111, 114, 123, 131, 222n80, 234n10; on hill town conditions, 100–101, 134, 136, 237n66, 238n1; recommendations of, 64–65, 117, 147; on towns similar to those studied, 66–67, 228n59. *See also* Land Utilization Committee

Thompson, Dorothy (*Concerning Vermont*), 185

Thompson, Zadock (*Gazetteer of the State of Vermont*), 19, 21, 29, 217n23

Thoreau, Henry David, 33–35

tourism, 65, 176–77, 178, 187, 207

towns: representation of in state government, 49, 66, 73, 226n25; elected and appointed offices, 164–66; work for, 166–67

Townshend, Vermont, 19, 136, 191, 197

Traditions and Ideals, Committee on, Vermont Commission on Country Life, 60–61, 175, 177, 230n21

"Vermont: A Way of Life" (Hard), 179–91

Vermont Commission on Country Life, 2, 8, 54, 56–57, 79, 175; 1929 meeting of, 57–61, 175; diagnoses of rural problems, 58–61; membership of, 57, 227n47, 231n22; relationship with Vermont Eugenics Survey, 61, 66, 71, 229n10; studies generated by, 2, 8, 54–55, 61, 213n2. *See also* Committees on Farm Production, Forestry, Human Factor, Land Utilization, Traditions and Ideals; *Rural Vermont: A Program for the Future*

Vermont Valley (Hard), 179

Wadman, Martha (eugenics investigator), 74

wages: in Cornwall and Waitsfield, 128; in Jamaica, 118, 127–28, 144; in larger towns, 142–43, 242n57, 242n59; for women, 128–29. *See also* incomes, standard of living

Waitsfield, Vermont: 5, 76; classification as "high type of town," 74; educational opportunities in, 137–38, 141–42; inheritance patterns, 140; as interpreted by investigators, 76, 85–86, 235n12; reasons for population persistence, 86–90, 94, 233n60, 233n63; types of farms, 108–9, 113, 119–23, 127–28

Wardsboro, Vermont, 67, 191, 228n59. *See also* Thirteen-Town Study
Warren, George F. (agricultural economist), 47
Warren, Vermont, 228n59. *See also* Thirteen-Town Study
We Americans (Anderson), 9, 73, 230n18
Wilson, Harold Fisher (*Hill Country of Northern New England*), 215n1, 218n39, 224n6
Winhall, Vermont, 14, 191–92, 205–6, 218n40
Wisconsin, University of, State College of Agriculture, 47, 57, 226n32
women: absences of in Migration Study, 92, 102, 128–29, 169, 242n73, 251n5; as eugenics investigators, 69–70, 80–81, 232n38; as workers in Jamaica, 128–29, 169–70; as workers studied by home economists, 129–30

Woodworth, Harry (agricultural economist), 194–95, 250n51
wool trade, 21–25, 219n44, 219n47, 219n51. *See also* sheep
work, off-farm, 39, 103–4, 113–18, 125, 144, 236n46; declining opportunities in Jamaica, 64, 100, 114–15, 136, 145; experts' disapproval of, 115–17, 171; in traditional skilled trades, 99–100, 114. *See also* ferning; logging; manufacturing; road work; town work

yields, crop, 30–32, 36; 221n68. *See also* buckwheat, cordwood, corn, cream, maple products, milk, potatoes
Young, Harry P., as chair of Land Utilization Committee, 58–60, 77; as dairy specialist at Vermont Agricultural Experiment Station, 149–150

www.ingramcontent.com/pod-product-compliance
Lightning Source LLC
Chambersburg PA
CBHW030531230426
43665CB00010B/841